P9-CAP-964

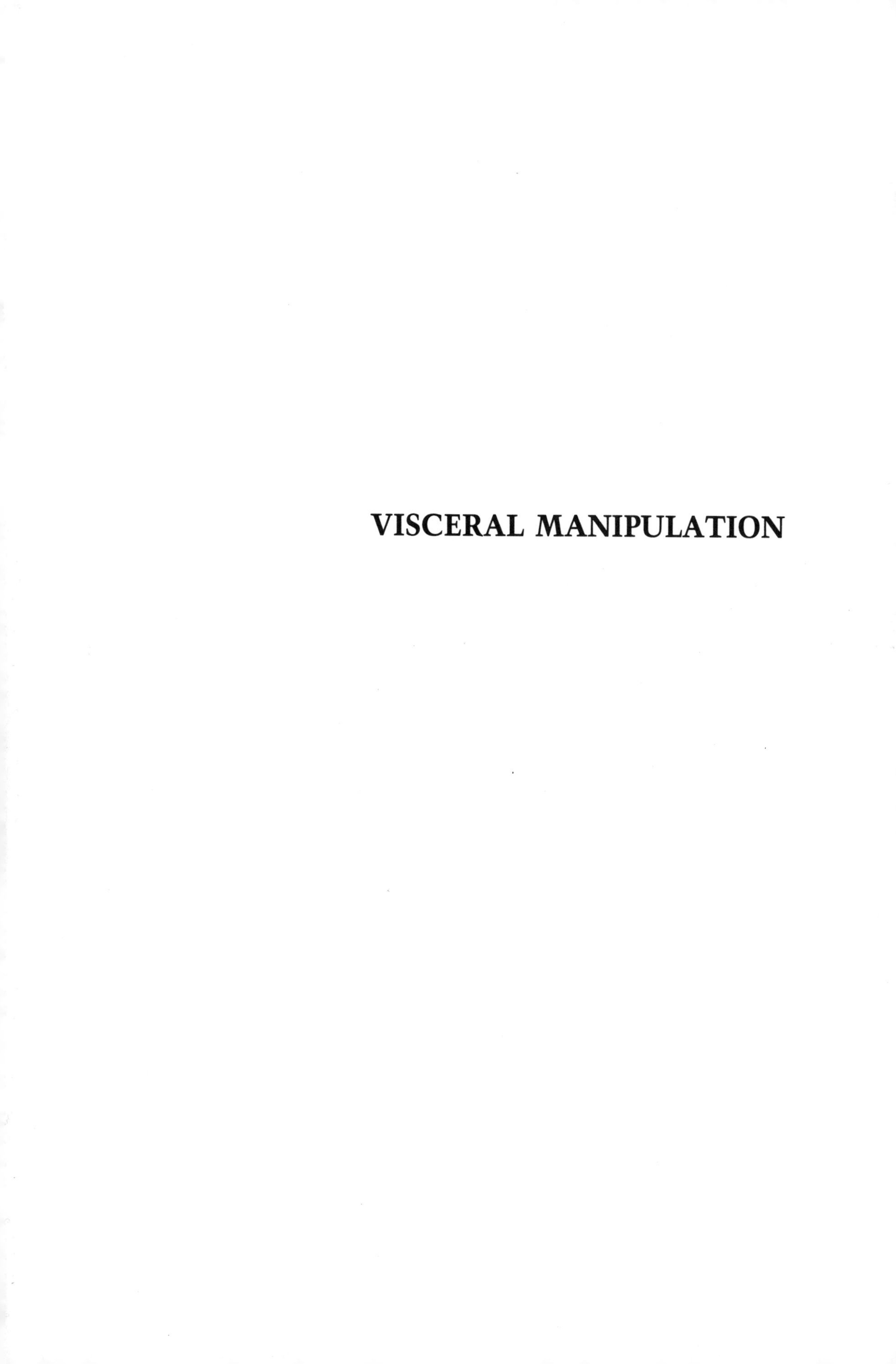

VISCERAL MANIPULATION

Visceral Manipulation

Jean-Pierre Barral
Pierre Mercier

ILLUSTRATIONS BY
Jacques Roth

Eastland Press
SEATTLE

Originally published as *Manipulations viscérales,*
Maloine (Paris), 1983.

Library of Congress Catalog Card Number: 92-82743
International Standard Book Number: 0-939616-06-8
Printed in the United States of America

Eleventh Printing, 2001

Book design by Catherine L. Nelson

Table of Contents

Foreword

I was honored and delighted when Messieurs Barral and Mercier asked me to write a foreword to their book.

I have known both authors since 1977, and have experienced their work personally. I have observed them, or worked with them, in the diagnosis of patients, and feel that I am reasonably familiar with their techniques. Both authors visited and spent time working with us at the Michigan State University College of Osteopathic Medicine during my tenure there.

The concept of visceral manipulation put forth in this book is a rather difficult one to explain fully in modern scientific terms. The clinical results, however, cannot be denied. Observation of these results requires the scientist to believe what is seen and to accept the fact that there are some human biological phenomena which are, as yet, inexplicable within the context of conventional medical science.

From my observations, it does indeed seem that the authors are able to detect an axis about which each visceral organ physiologically rotates. It appears that they are also able to detect an improperly directed axis, and, more astonishingly, to effect change in the function of the viscera in question by correcting the direction of its axis.

The concept that each viscera has an inherent motion which is intimately connected with the physiological functions of that viscera is fascinating. The fact that this motion can be modified by manual means is truly significant. The potential for application of this concept in the diagnosis and treatment of internal organ dysfunction is, in my own opinion, limitless. Its use in preventive medicine and health care could be one of the most important contributions of our life time.

John E. Upledger, D.O., F.A.A.O
President
The Upledger Institute
Palm Beach Gardens, Florida

Preface

Nature abhors a vacuum, but fears immobility even more. Motion is a sign of life itself. Primordial energy, the first mover, is at the origin of all form and all formulation. From the infinitely large to the infinitesimally small, life is always in motion. It all began with an impetus, a vibration. Everything in the universe is in motion, whether of large or small amplitude, of high or low velocity. Electrons dance with an unbridled rapidity, while the displacement of the tectonic plates of the earth's crust is but a few centimeters a year. Everything moves in space and in time, and humans are no exception to this rule.

Humans are both an integral part of the cosmos and also whole unto themselves. We are composed of bony articulations, muscles which allow movement, and visceral structures which assure the function of the whole. Life is movement, rhythm, exchange and perpetual adaptation to new situations, assimilation and rejection, and defense. This is continuous up to the time of death, when all seems to come to a stop.

The vertebral column is an extremely important component of the body. Its resilience is due to the fact that it is a flexible and deformable structure. As we shall see, the healthy physiology of the viscera also depends on this capability of being deformed. The viscera of the abdominal cavity move freely with respect to each other because their enveloping serous membranes make up sliding surfaces. The groups of mobile viscera are contained in the abdominal, pelvic, thoracic and cranial cavities.

All pathology of the viscera results in what we will call visceral restrictions. When this happens, the viscera in question is no longer freely mobile in its cavity but is fixed to another structure. The body is forced to compensate for this situation, which leads to a functional problem and eventually, if the compensation is inadequate, to a structural problem. This book will point out the pathological findings of visceral restrictions. Treatment then consists of stimulation of the viscera in order to restore its primary physiological mobility and motility.

A structural theory of disease and treatment cannot be sufficient in itself to explain and treat all pathological phenomena. This is especially true if the theory is limited to the vertebral system to the exclusion of all other systems. The concept that all pathology and therapeutics can be encompassed by the vertebral reflex arcs, or (in the most reductionist view) by the first two cervical segments exclusively, has not contributed to the credibility of our art.

Osteopathy seems to be divided into two schools, the mechanistic and the energetic. For those adept at the mechanistic theory, the energetic theory appears to be just so much "magnetic charlatanism," while for proponents of the latter, the direct action manipulators appear to be "muscular savages." But in fact, osteopathy must be an integrated whole, utilizing both approaches.

The energy theory is based on the idea that people always produce, gain and lose energy. When one is in good health, these energy exchanges take place in a balanced and harmonious manner. When one is in poor health, however, the balance of energy, local or systemic, is disrupted. These energy exchanges take place in our interior milieu, as well as in our relationships with the universe. A human is but a tiny link in the great ensemble of cosmic energy.

Osteopathy must study and describe all of these movements and exchanges, from the smallest to the greatest, analyzing their perturbations, in order to be able to effect any adjustment or correction. While some movements are readily visible, others are not, because of their rapidity and/or small amplitude. For example, all we can perceive with the naked eye is a muscular contraction; the thousands of tiny cellular movements involved can be seen only with the aid of a microscope. We must remember that this multitude of tiny movements is what makes up the whole.

Osteopathy should be concerned with all that moves in the human body, from the smallest and simplest movements to the most complex. This leads to a confluence of the structural (mechanistic) and energetic approaches. Osteopathic treatment, whatever form it takes, is an energetic action because the fact that we can have an impact on motion contributes to an improvement in the distribution of energy. All human body systems can be stimulated, inhibited or changed by use of the hands. Dexterity is required for all types of manipulation, including those affecting the visceral system as well as those affecting the vertebral column or the structures of the craniosacral system. This dexterity, combined with an understanding of body function and structure, is what makes a true osteopath.

Osteopathy is the art of provoking self-correction on the part of the organism, and visceral manipulation is only one of the means (though an important one) to this end. Isolated manipulations of the sacroiliac joint, sphenofrontal suture or liver are only of relative interest and should never be seen as ends in themselves. Each is a means of gaining access to the whole system and thus calling forth a self-corrective response. Osteopathic manipulation stimulates the organism to mount its own defenses and to draw from its own reserves; the manipulation is no substitute for the body's own power. With the above in mind, we offer this book for your consideration.

There are many people we would like to thank for helping us understand the osteopathic concept and enabling us to forge ahead with the concepts discussed in this book. Our primary teachers have been Ange Castejon, D.O. and Thomas J. Dummer, D.O. who were respectively director of the French branch and principal of the European School of Osteopathy (EEO) in England during our tenure there. John S.G. Whernam, D.O. was the previous director of the EEO and a student of John Littlejohn, who was himself a disciple of Andre Taylor Still. The work of Irwin Korr, Ph.D. throughout his long career has had a great influence on osteopaths throughout the world. We have also been influenced by the spirit and work of John E. Upledger, D.O., F.A.A.O., especially that performed while he was a faculty member of the Michigan State University College of Osteopathic Medicine.

Denise Gilles, Paulette Mercier and Genevieve Planchard worked with our ideas and prose to help fashion the original French edition of this book. We would particularly like to thank Stephen Anderson, Ph.D. and Daniel Bensky, D.O. of Eastland Press for all the time and effort they put into helping us revise this book for an English-speaking audience. With their help, we believe that we have improved upon the original.

Chapter One: Basic Concepts

Table of Contents

CHAPTER ONE

Basic Concepts

The hypothesis put forward in this work is that an organ or viscera in good health has physiologic motion. This motion is an interdependent one because of the serous membranes which envelop the organ, and the fasciae, ligaments and other living tissues which bind it to the rest of the organism. Physiologic motion can be divided into two components: (1) visceral mobility (movement of the viscera in response to voluntary movement, or to movement of the diaphragm in respiration); and (2) visceral motility (inherent motion of the viscera themselves). All viscera should function properly, without any restrictions. Any restriction, fixation or adhesion to another structure, no matter how small, implies functional impairment of the organ. The consequent modification of its motion, repeated thousands of times daily in the body, can bring about significant changes, both to the organ itself and to any related structures. It is our experience that through manipulation it is possible to bring about an improvement in function by restoring some measure of proper motion.

In this chapter we will focus on the physiology and pathology of both types of visceral motion, as well as the experiments which have permitted their study. A new concept, that of visceral articulations, has emerged from this type of study; we will look briefly at the physical laws which govern these articulations.

The Different Motions

Visceral motion can be divided into four categories, according to the system which influences or controls them:

- somatic nervous system
- autonomic nervous system
- craniosacral rhythm
- visceral motility

MOTION INFLUENCED BY THE SOMATIC NERVOUS SYSTEM

Motion controlled by the somatic nervous system is the easiest to observe. It is part of the motor system, which governs all voluntary movement, and has been the subject of extensive studies. It is readily observed in terms of gait, movement of the trunk, etc., and has been systematized according to the anatomy and physiology of striated muscle. Voluntary motion is the result of mobilization of skeletal structures under the control of the central nervous system. It is important to remember that observable gross movement is the consummation of numerous small movements, often involving multiple articulations.

The motor system is a source of passive motion of the viscera and, as such, is a factor which enters into visceral mobility. The viscera are passively moved by walking, running, flexion of the trunk and other gross skeletal movements. All the viscera are contained within one of three cavities — the skull, thorax, or abdomen. The latter two cavities are only partially formed by the skeleton, and can easily be deformed (along with their visceral contents) by bodily movements. The anatomical relationships of contiguous viscera vary, depending on changes in voluntary skeletal motion. Knowledge of visceral anatomy and of the physiology of skeletal motion enables us to predict the direction and amplitude of visceral motion.

For example, if a subject stands and bends forward at the waist, his liver will move forward, sliding over the duodenum and the hepatic flexure of the colon below. The liver and the hepatic flexure will both move inferiorly, but the liver more so, since it moves first and farthest with flexion. Thus we can say that the liver slides anteroinferiorly over the duodenum and hepatic flexure, even if these other structures move in the same direction. Similar processes occur in the other viscera, as will be discussed in detail in the appropriate chapters below.

MOTION INFLUENCED BY THE AUTONOMIC NERVOUS SYSTEM

Gross, voluntary motion is dependent on the somatic nervous system. The autonomic, vegetative functions are controlled, to varying degrees, by the autonomic nervous system, as well as the endocrine system. Autonomic movement, which has a direct or indirect impact on the viscera, include diaphragmatic motion, cardiac motion and peristaltic motion.

Diaphragmatic Motion

While diaphragmatic motion has been well-described as regards the mechanism of respiration, its action on the abdominal viscera has been neglected by physiologists. The diaphragm completes approximately 24,000 movements a day, pulling and pushing the lungs and abdominal viscera along with it each time.

The trunk contains the pleural and peritoneal cavities, which are anatomically closed and structurally related by their contiguity. Through the diaphragm, which separates them, they are also functionally related. The diaphragm acts like a piston, rising and falling in the cylinder formed by the trunk. During inhalation, when the diaphragm descends, it produces expansion of the thorax and compression of the abdomen. This is a simplified model, as the trunk is not actually a rigid cylinder, nor is the diaphragm a flat piston.

As the diaphragm drops during inhalation, there is a decrease in intrathoracic pressure which, in turn, causes air to flow into the alveoli by way of the trachea and bronchioles. The volume of the thoracic cavity is thus increased.

What happens to the abdominal cavity when the diaphragm descends? The total volume of the abdominal viscera is incompressible, and the residual space between organs is minimal. In order to allow for the descent of the diaphragm, the abdominal cylinder will have to deform. Posteriorly and inferiorly, the cylinder is composed of skeletal structures: the vertebral column and pelvic girdle. The force exerted by the movement of the diaphragm is insufficient to deform these structures, but it is sufficient to push the anterior abdominal wall, composed only of muscle and connective tissue, forward. The volume lost due to a shortening of the distance between the diaphragm and the pelvis is regained by increasing the anteroposterior diameter.

The continual deformation of the abdominal wall, fluctuating between the two extremes of end-inhalation and end-exhalation, will cause the viscera to slide and rub against one another within the abdomen. The pressure of the diaphragm is vertical (directed downward), but has a resultant horizontal force against the anterior abdominal wall (see chapter 3). Knowing the direction of these forces allows us to determine the direction of motion of each viscera during respiratory movements. The mechanism of these forces is extremely complex, for there are no flat surfaces moving against each other. There are rather forces (ascending, descending, oblique and circular) which are reflected and rebound according to the positions and relationships of the surrounding skeletal and soft tissue structures. An organ does not move on one plane only, but on three planes: sagittal, frontal and transverse.

Visceral mobility, although passive, does exist and is quantitatively very important. Remember that the diaphragmatic pump moves 24,000 times a day. Variations in pressure can easily lead, in a pathological state, to deterioration of the structures mobilized by the diaphragm.

Cardiac Motion

This motion is repeated approximately 120,000 times a day, and directly affects the lungs, esophagus, mediastinum and diaphragm. The diaphragm transmits these vibrations to the abdominal cavity together with its own rhythmic movement. The wave motion of blood leaving the left ventricle is propagated via the arterial bed to the farthest capillary of the most distant organ. Thus, even the smallest restriction can take on considerable importance as it is stressed 120,000 times a day along a pathologically modified axis.

Peristaltic Motion

Peristaltic motion consists of great contractile waves, stirring and circulating the visceral contents. It involves the hollow organs and is under the influence of neuronal, chemical and hormonal factors. It has less effect on visceral mobility than do the diaphragmatic and cardiac motions.

CRANIOSACRAL RHYTHM

The structures of the central nervous system (i.e., the brain and spinal cord) are bathed in cerebrospinal fluid (CSF). The CSF is not stagnant; it is in constant motion, circulating under the influence of the craniosacral rhythm.

This rhythm consists of two movements: flexion, which is active, and extension, which is passive. On flexion (or expansion), there is a decrease in the anteroposterior dimensions of the skull and body, accompanied by an increase in width. On extension (or relaxation), the opposite occurs; the head and body become narrower. In terms of the motion of the paired or bilateral structures, flexion and extension are manifested as external and internal rotation, respectively.

The craniosacral rhythm and its importance were discovered and first described by William G. Sutherland, D.O., over fifty years ago. Sutherland referred to the phenomenon as "primary respiratory motion." The work of John E. Upledger, D.O., while he was in the Department of Biomechanics at Michigan State University, led to what we consider at this time to be the most plausible explanation of this rhythm.

According to Sutherland, a fluctuation in the flow of CSF affects the bones of the skull, as well as other skeletal structures. This results in motion of the cranial bones. This concept differs from previous textbook descriptions of the skull as an immobile unit, with sutures that are closed very early. In Upledger's scheme, the craniosacral rhythm results from variations in pressure among arterial blood, CSF and venous blood. In effect, the CSF is filtered and diffused out of the arterial system, making its way into the subdural space and the arachnoid villi before diffusing back into the blood stream via the venous system. Upledger's contribution is the recognition that the formation of CSF is periodic and rhythmical, not continuous. During the flexion phase, CSF is secreted into the brain, and there is an expansion of the semiclosed hydraulic craniosacral system during which the ventricles swell and all horizontal diameters are increased. Once this expansion reaches a certain threshold, receptors in the skull sutures that are sensitive to stretch reflexively stop the secretion of CSF. Because the reabsorption of CSF into the venous system is continuous, the pressure decreases and the extension (or relaxation) phase begins. As the CSF pressure in the ventricles drops, the sutures compress because CSF volume within the craniosacral system is reduced. CSF production is turned on again as a new flexion phase follows the extension phase *(Upledger 1983).*

The craniosacral "pump" is a passive one, resulting only from the circulation of CSF between high and low pressure areas, under the influence of the baro- and mechanoreceptors. It would be enlightening to see a graph representing the pressure variations of CSF and venous blood. The difficulty, however, is due to the small magnitude of these pressures: CSF pressure is between 12-15cm H_2O and venous pressure 5-10cm H_2O. Arterial pressure is much higher, around 120cm H_2O.

There is currently no adequate explanation for the frequency of the craniosacral rhythm, which is on the order of 8-12 cycles per minute. We know that it is not influenced by diaphragmatic respiration, cardiac rhythm or the activity of the individual. The rhythm is not under voluntary control; it is one of the body's autonomic systems. It is possible to conceive of this automaticity as being controlled by an archaic or primordial brain according to mysterious genetic laws. Some researchers believe that cells have both a spatial and a temporal organization, such that each cell has a memory and is programmed to begin a cycle under the influence of unknown factors.

VISCERAL MOTILITY

With the exception of peristalsis, all of the visceral movements described above are passive and influenced by extrinsic factors. But the viscera also have an intrinsic, active motion which we call motility. They move independently, with a motion which is slow and of such low amplitude as to be almost imperceptible. Visceral motility is perceptible to the hand but requires an educated sense of touch. It is the kinetic expression of tissues in motion. We have no scientific explanation for this phenomenon, and are aware of it only from experience. Is it just an extension of the craniosacral rhythm, or does it correspond to movements of the organs during embryogeny?

During embryogeny, a series of cellular modifications take place, beginning with a fertilized egg and culminating in the complex fetal organism. Cellular development is not anarchic, but proceeds according to a well-defined order in space and time. There is a "coordinator" responsible for the harmonious alignment and development of cells and tissues. The cell has a memory, based at least in part on the molecules (DNA, RNA) which carry genetic information. In the words of Rollin Becker, D.O.: "The tissues alone know." Organs migrate during embryogeny. For example, the stomach rotates to the right in the transverse plane and clockwise in the frontal plane. The transverse rotation orients the anterior lesser curvature to the right, and the posterior greater curvature to the left. The frontal rotation moves the pylorus superiorly and the cardia inferiorly.

The embryologic theory of visceral motility postulates that the axes and directions of these motions remain inscribed in the visceral tissues. Thus, visceral motility occurs around a point of equilibrium, oscillating between an accentuation of the embryologic motion and a return to the original position, with a contractility analogous to (but much slower than) that of the nodal tissue of the heart.

The motility cycle has two phases, in which the organs move toward and away from the median axis of the body. We call these phases "expir" and "inspir" respectively, using neologisms in order to avoid any confusion between these movements and those of diaphragmatic inhalation and exhalation, or flexion and extension of the craniosacral system. Under normal conditions the organs move in sync, i.e., they all undergo inspir or expir at the same time. Note that there is no particular relationship between the direction of motion of the organs during the different phases of visceral mobility and those in the phases of visceral motility. In some cases (e.g., the liver), the motion of an organ in inhalation is similar to its motion in expir, while in other cases (e.g., the kidneys), it is similar to its motion in inspir. For still other viscera (e.g., the colon), the motions of mobility and motility are completely different.

We do believe that there is a relationship between visceral motility and the craniosacral rhythm, although we do not at present know exactly what it is. We have defined the terms inspir and expir so that they coincide with craniosacral flexion and extension. Therefore, they can be quite different from the motions of the organs in response to pulmonary respiration. For example, the liver in inspir rotates posterosuperiorly, as do the hypochondriac regions during the flexion/external rotation phase of the craniosacral rhythm. This is almost the exact opposite of the anteroinferior motion of the liver during the inhalation phase of pulmonary respiration.

Another similarity between the phases of these two inherent motions is that inspir is relatively more active than expir (as is flexion compared to extension). Many of the motility tests described in the following chapters will concentrate on following the expir phase, because there is less resistance in this phase. The relatively decreased resistance makes following the motion testing easier, and therefore releases are easier to obtain. When following inspir, if you do not follow in precisely the correct direction, the motion will stop without leading to a release.

An interesting confluence of these two inherent motions occurs in the head. Here, in addition to the craniosacral rhythm, the visceral motility of the brain can be felt. It is much more of a straightforward forward bending (expir) and backward bending (inspir) rotational motion than the craniosacral flexion and extension phases, which also contain an expansionary or external/internal rotational component. We have not yet investigated the clinical utility and significance of this motion, except as a means of releasing tensions in the dura mater.

Amplitude is an essential parameter of all motility. It varies from organ to organ (e.g., the liver normally has a larger amplitude than the ascending colon), and will decrease because of reduced function of an organ, or restrictions of the surrounding connective tissues. The quality of the motion is at least as important as its amplitude (quantity), and treatment should be focused on both parameters.

RHYTHMS

The diaphragmatic respiratory rhythm is usually of the order of 15-18 cycles per minute and can be voluntarily modified. The craniosacral rhythm is of the order of 8-12 cycles per minute and is influenced to a much lesser extent by external phenomena. The visceral rhythm is characteristically 7-8 cycles per minute. The craniosacral and visceral rhythms may be reduced if the subject is ill or tired. Craniosacral rhythm and visceral motility are similar in that they are both inherent motions of the tissues and their frequency is not affected by outside influences. As mentioned above, the relationship of these rhythms to each other remains to be understood.

Peristalsis varies in response to local and systemic factors. It is discontinuous, with long periods of rest characteristic of each of the viscera. Thus, a full stomach will produce peristaltic waves every 3 minutes, requiring 20 seconds to traverse its full length. It is important not to confuse this peristaltic rhythm with visceral motility.

Research

EMBRYOLOGIC AXES

Our research has been, from the beginning, strongly rooted in our clinical experience. The movements of motility and mobility have been studied for each organ. There is an inherent axis of rotation in each of these motions. In healthy organs, the axes of mobility and motility are generally the same. With disease, they are often at variance with one another, as certain restrictions affect one motion more than the other. What a surprise it was for us to discover that the axes of motion reproduce exactly those of embryological development! Neither preconceived ideas nor hypotheses directed this research. The discovery of this phenomenon was purely empirical, and tends to confirm the idea that "cells do not forget."

MOBILITY

The mobility of the viscera under the influence of the piston-like movement of the diaphragm is a familiar phenomenon, easily documented by simple radiology. Our work began by looking at pulmonary pathology and its effect on other viscera. We had an opportunity to observe and participate in dozens of cases in a pulmonary disease ward in which many types of infectious lung diseases were treated (our thanks to Drs. Arnaud and Roulet of Grenoble, France, who collaborated with us on this research). We saw the progression of diseases as well as their sequelae and were, in some cases, able to participate in the post-mortem. Many of the patients had been previously treated by therapeutic pneumothorax.

Our observations demonstrated that whenever the pleuropulmonary unit is attacked, the axes of motion of the thoracic cavity and viscera change. The directional forces of intrathoracic pressure are changed, and all the musculoskeletal structures related to the thorax then move along different axes. These changes have many repercussions. The gastroesophageal axis is deviated greatly, increasing the risk of hiatus hernia. The rotation of the stomach is altered by transitory perturbations. The positions of the heart and pericardium are altered. The pleuropulmonary attachments pull on the lower part of the cervical spine. Costovertebral articulations lose their elasticity, and so on. In the end, the whole body is involved. These disordered changes are visible and easily palpable. We were able to observe the changes in elasticity to which the connective tissues were subjected. Some tissues doubled or tripled in thickness in adapting to abnormal tension. When the possibility of adaptation was exhausted, fibrosis then set in. In the phenomenon of spontaneous pneumothorax, it appeared that the pleura had been subjected to previous abnormal tensions (e.g., violent arm motion) or weakening secondary to septic microphenomena. Adaptive scoliosis following some thoracic surgeries or therapeutic pneumothorax is a familiar phenomenon. This suggests that the forces generated by changing visceral mobility are significant and, over time, able to grossly deform the skeletal structures.

In support of these observations, we also noted that a relatively minor pleural injury could be responsible for a considerable pathology which, in turn, could lead to other disturbances, such as chronic cervical neuralgias. A small disturbance in motion, repeated millions of times over months or years, can provoke problems seemingly disproportionate to the original cause. This illustrates the law of geometric progression: from minor causes flow major effects, which can be at some distance from the source of provocation. For example, the kidney moves 3cm with each breath, which cumulatively amounts to 600 meters a day! With extremely forced respiration it will move as far as 10cm. A minor disturbance in the response of the kidney to breathing can therefore cause a major problem over time.

MOTILITY

Remember that motility is inherent to and characteristic of each organ. It is the motion which can be felt when all extrinsic influences, particularly that of diaphragmatic motion, have been screened out. While we believe that the existence of the craniosacral rhythm has been demonstrated, the motility of the viscera has not yet been clearly proven. We have made use of ultrasound and fluoroscopy with enhancement in order to try to prove its existence. Real-time ultrasound permits the observation of an organ without freezing time or any risk to the patient or operator. The effect of the pumping of the heart is seen more as a vibration than an actual movement of the organ. One can easily perceive the motion of entire organs, but it is difficult to precisely describe motility and define the axes. In an attempt to isolate motility, we first asked our subjects to hold their breath. An apneic state is obviously not physiologically normal, since it brings about a contraction of the abdominal muscles and a consequent elevation of abdominal and thoracic pressure. The result of these pressure changes is an inhibition of motility; we were unable to observe it under these conditions.

In order to perceive motion in an apneic subject, it would seem better if the breath could be held with the lungs "half full." In this position, the intrathoracic and abdominal pressures are balanced. However, we must admit that certain motions which we have been able to observe as independent of diaphragmatic motion could not be reproduced reliably under these conditions. Possibly the stress of holding one's breath was responsible for this failure to obtain reproducible results.

As an initial step, we have at least been able to focus on these movements a few times. The most distinct movements were observed with enhancement using intravenous pyelography (IVP) or cholecystography. The clearest event which we were able to observe was in a young man undergoing IVP. He presented with a condition in which his kidneys dropped about 6cm while he was in a standing position, which meant that they were very loosely attached to the contiguous structures. As he held his breath, his kidneys continued to move, repeatedly, with an amplitude of 3cm in the vertical and lateral directions. We are continuing in this line of research, thanks to the extreme kindness of radiologist Serge Cohen, M.D.

The only means of proving motility at this time is to focus on the reproducibility of palpatory findings. The criteria for reproducibility are that:

- several practitioners perceive the same thing, on the same subject, without previous knowledge or collusion, and this result is obtained several times
- a technique produces the same results on different subjects

Many of our treatments based on visceral motility have met these criteria.

The motility of an organ is affected in a few ways. Restrictions in the surrounding tissues can cause adhesions or fixations which change the axes, upset the symmetry, and decrease the amplitude of motility (see pages 17-18). In hollow organs, local viscerospasms can also affect the axes and amplitude of motion (see pages 19-20). In addition, with infectious processes, their sequelae or other processes that affect the parenchyma of an organ, the amplitude of motility decreases drastically and often becomes fixed in expir. Examples of this are pneumonia, hepatitis, cirrhosis and nephritis.

DIFFERENT CYCLES

Ancient Oriental health practitioners realized that people are subject to external influences, some of which are cyclical, that are capable of changing behavior and function. According to Oriental medical theory, energy circulates throughout the body and reaches its zenith in different organ systems or acupuncture channels at specific times *(Figure 1)*. It is strongest in the lungs between 3:00 and 5:00 AM, passes into the large intestine between 5:00 and 7:00, then into the stomach, and so on. The circuit terminates at the liver between 1:00 and 3:00 AM. Each organ, during its particular hour, is at the height of activity.

We have observed evidence for such a cycle. The time at which an organ is at its zenith is expressed not by an acceleration of rhythm, but by an increase in the vitality and amplitude of its motility. Our studies in this area are incomplete, as we cannot measure the motility of the organs we cannot palpate.

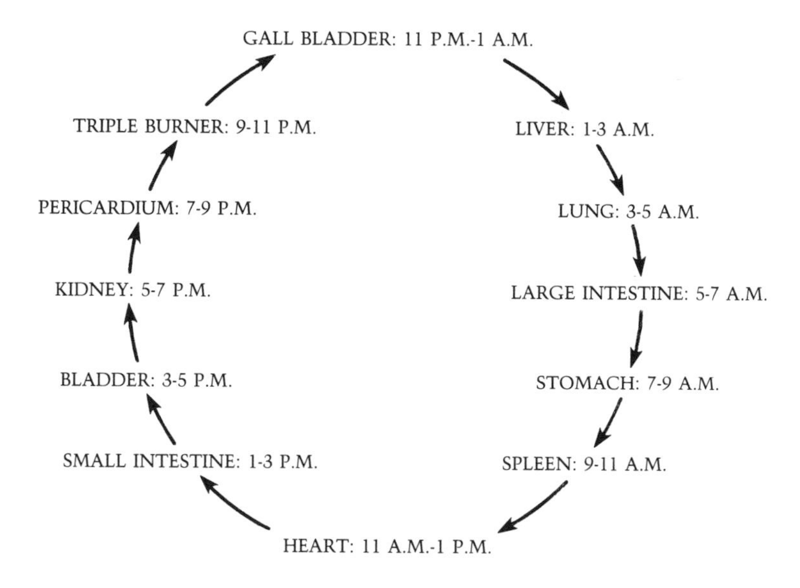

Figure 1
Oriental Medical Circulation of Energy Through Organs

There are other cycles (e.g., seasonal and annual ones) that also affect humans. The duration of certain cycles' stages, such as puberty or menopause, may be quite long. All of these cycles can be important.

Scientists have studied daily variations in cortisol level. They have observed a nocturnal increase in parasympathetic tone, as evidenced by visceral discomfort. Laboratory studies have also demonstrated five peaks per day in the circadian rhythm of catecholamines. There are many unanswered questions in this area. Which clock regulates these rhythms? What causes an individual to die at a precise moment? What sets off labor in a pregnant woman? And, as Voltaire would have asked, if there is such a clock, must there also be a clockmaker?

The cycles that we are most concerned with here, pulmonary respiration (as it affects visceral mobility) and visceral motility, are integrally related to all the other cycles that affect humans. In our studies, we must always be aware of the effects of these other cycles. They are all manifestations of the primordial importance of motion.

Visceral Articulations

Skeletal motion results from voluntary muscular activity originating in the central nervous system. Muscles mobilize contiguous skeletal structures, which are unified by a system of articulations. The nature of these articulations determines the axes and the amplitudes of voluntary movements.

There are also defined axes and amplitudes for the motion of visceral structures, which leads us to the concept of visceral articulations. Visceral articulations have, in common with skeletal articulations, sliding surfaces and a system of attachment. The difference lies in the absence of muscle attachments to supply the moving force.

SLIDING SURFACES

The sliding surfaces of visceral joints are the serous membranes. An organ can be contiguous with a muscular wall (liver-diaphragm), with the skeleton (lung-thorax) or with another organ (liver-kidney). These serous membranes are called:

- meninges (central nervous system)
- pleura (lungs)
- peritoneum (abdominal cavity)
- pericardium (heart)

Viscera, which undergo changes in volume and move with respect to neighboring organs, are covered by a fine visceral membrane, consisting of a layer of flat mesothelial cells and an underlying layer of dense connective tissue. In the thorax this membrane is known as the visceral pleura; in the abdomen as the visceral peritoneum. There are lines of reflexion (e.g., at the hilum) where the visceral membrane blends with the parietal membrane lining the walls of the cavity. Between the visceral and parietal membranes, there is a small quantity of serous fluid which functions as a lubricant and as part of the immunological system. If the membrane's function in visceral articulations is analogous to that of articular cartilage in synovial joints, the serous fluid acts like synovial fluid. Thus, all viscera are covered with a membrane which is in contact only with another membrane, and the contact is mediated by a natural lubricating fluid.

ATTACHMENTS

In the three cavities of the trunk (pleural, pericardial and peritoneal), the various organs are held in place by very different means, and the points of attachment are multiple and varied. The systems are:

- double layer system
- ligamentous system
- turgor and intracavitary pressure
- mesenteric system
- omental system

The order in which they are listed corresponds to the relative importance of their roles in sustaining the attachments.

Double Layer System

We have already touched on the importance of this system, in which serous membranes and serous fluid have a function analogous to that of cartilaginous surfaces and synovial fluid. The double layer system, however, also creates a suction effect. The two layers which lie flat against each other, separated only by a thin liquid film, follow the law of relative pressures and cannot physiologically separate themselves. The only motion which is possible is a sliding over each other, as in the model of two pieces of glass separated only by a layer of water. This double layer system is operative in pulmonary, cardiac, peritoneal and cerebrospinal visceral articulations.

Ligamentous System

The ligaments which bind viscera are different from those which bind skeletal structures. When thinking about viscera, forget the image of ligamentous stability as a last stronghold preventing subluxation. Ligaments of the abdominal and pulmonary viscera are folds of pleura or peritoneum binding an organ to the wall of the cavity, or binding two organs together. Their role is to maintain the viscera in their proper place. They act as constraints against gravity during voluntary and respiratory movements; i.e., their role is one of containment. They are generally avascular but have a rich nerve supply. A few ligaments are important in suspending viscera. Primary examples are the ligaments of the pleural dome in their constant struggle against the traction of the diaphragm and the weight of the suspended viscera, and the coronary ligament which is the indispensable link between the diaphragm, liver and stomach.

Turgor and Intracavitary Pressure

Turgor and intracavitary pressure are essential factors in holding the viscera in place. One can consider the volume of the abdominal cavity as constant. But, depending on their activity, individual viscera can change in volume. This is possible because of turgor, which we define as the sticking together of the organs due to their swelling in a discrete space.

Turgor acts upon the hollow intraperitoneal organs. Each hollow organ, under the effect of its elasticity, vascular system and contents, occupies the maximal amount of space made available for it. This phenomenon enables the viscera to remain constant in mass and yet to stick together. The intravisceral gaseous pressure of hollow organs accentuates this cohesive effect. The net effect of turgor is to increase the cohesion between viscera by means of gaseous and vascular tension. This explains why the extravisceral volume in the abdominal and thoracic cavities is so small, requiring such a small amount of serous fluid. Anything which contributes to dilating the lumen of an organ is a factor in cohesion of the visceral mass.

Intracavitary pressure is equal to the sum of the pressures within each organ (intravisceral pressures), plus the pressure between the organs. It is another means by which the abdominal organs are pushed together. Intracavitary pressure must always be in equilibrium with extracavitary pressures: weight, atmospheric pressure and the tension of the muscular wall. The combination of turgor and intracavitary pressure allows for the maintenance of a relatively homogenous visceral column upon which the diaphragm presses during inhalation.

The trunk consists of two main cavities, thoracic and abdominal, which are separated by the diaphragm, and which enclose the lungs and abdominal viscera respectively. The problem of pressures will be taken up in chapter 3, which deals with the abdomen. Here we will simply mention a few essential points, particularly the fact that one of these cavities is at a much lower pressure than the other, and that they must coexist harmoniously nevertheless.

The supradiaphragmatic pressure is lower than the subdiaphragmatic pressure, so that the contents of the abdominal cavity are in effect suspended from the diaphragm. The force of gravity in the abdominal cavity acts against this upward force. The organs which are closest to the diaphragm are most sensitive to this effect, and the further we descend into the abdomen, the more it is attenuated. The weight of the liver, for example, which is approximately 2kg, is effectively diminished by over 1kg by the attraction effect created by pressure differentials. This phenomenon explains why a tear or opening in the diaphragm always results in an upward migration of abdominal organs.

Mesenteric System

The mesentery exists only in the abdominal cavity. It consists of folds of peritoneum which are very relaxed. Its role of containment is minimal. There is a complex network of vessels and nerves between the folds reflecting their nutritive function. The mesentery is involved only with organs of the digestive tract (small intestine and colon), and attaches them to the peritoneum.

Omental System

The omenta, like the mesentery, are found only in the abdominal cavity and consist of folds of peritoneum which join two elements of the digestive tract. Being reinforced laterally, they play a minor role in the suspension of the organs to which they are attached, as we shall see later. Their neurovascular role is relatively important.

We have spoken of the physiology of motion and, in particular, of the different anatomical structures which facilitate and control visceral mobility. In the following chapters, we shall describe the parameters in particular viscera which regulate these motions. Each viscera has a specific function, and, in order to carry it out correctly, must be in good working order. The extrinsic motions which we have described provide healthy adaptations to the constraints imposed by motor and autonomic activities. Correct motility or intrinsic motion is a manifestation of the good health of the organ, in which it is able to function at its best. On the other hand, all obstacles to visceral motion predispose the organ to an abnormal physiology translating into functional difficulty.

If the general adaptation is inadequate, a structural lesion will result in which the fibers of the organ are modified, and treatment must take this into account. To borrow an analogy from Jacques Descotes, former President of the French Osteopathic Association (AFDO), the body is like a tightrope walker, making small adjustments right and left, backwards and forwards, in order to keep his balance. A small deviation from equilibrium can add up over time to a functional difficulty. Good health is not a state, but a search for equilibrium. The treatment must be light and subtle. If the tightrope walker should fall (i.e., a serious structural difficulty develops), a more energetic therapeutic intervention would be required.

Pathology of Motion

In the preceding section, we discussed the normal physiology of motion. Each organ moves in particular directions along particular axes. Changes in these motions may result from a variation of the axis or the amplitude, and may involve the inherent motility of the organ itself or the mobility determined by the structures of visceral articulation. Such changes will show one or more of the following:

- a clear local pathology, with symptoms
- the beginning of a local pathology, without symptoms
- a local sequela to an old pathology, to which the subject is well adapted
- a pathology at a distance in a viscera having "articular" relations
- a pathology in a structure having vascular, nervous or fascial relations

Because of the relationships of organs and systems within the body, we sometimes observe veritable chains of lesions. These chains can be made up of any combination of links between viscera, muscle, fascia and bone. The peritoneum is a link between all the viscera and related structures, and acts like a reciprocal tension membrane. All pathology causes perturbations in the motility of the affected organ. In order to distinguish these from pathologies of the musculoskeletal system, we call these perturbations *visceral restrictions.*

VISCERAL RESTRICTIONS

In this book, we use the word "restriction" to refer to any decrease in motion. A visceral restriction occurs when an organ loses part or all of its ability to move. We have discussed the fact that visceral articulations are made up of sliding surfaces and means of connection. A restriction may arise either at the level of these structures or on the walls of the actual organ, and can usually be assigned to one of three categories which we will discuss below: articular, ligamentous or muscular.

One can also distinguish functional from positional restrictions. With *functional restrictions,* only the function of the related organs is affected; their positional relationships are not changed. With *positional restrictions,* the anatomical relationships of the organs are changed and their articulations are modified. For example, with a right renal ptosis, the kidney loses all contact with the liver — a veritable visceral subluxation. Although uncommon, it is possible to have a positional restriction without a functional restriction.

Foodstuffs may cause restrictions of certain viscera. We have tested many of our patients, by having them either eat the food or just hold it in their hands. In the latter case, neither we nor the patient knew what the substance was during the procedure. We discovered that, in certain cases, a food substance can bring the motility of a viscera to an immediate halt. We were surprised to find that, apart from the well-known problem substances such as chocolate and white wine, many other foods also cause sensitivities. Remember, a substance may even increase the vitality and amplitude of the motility of one organ, while simultaneously decreasing the motility of another organ. It is the interaction of the substance and the patient which is important.

Many endocrinological, chemical, environmental and psychological phenomena can affect visceral motility. One example may be of interest. In almost every case of nervous depression, the motility of the liver is affected. The liver, in Oriental medical theory, is called the "sea of emotions" and is considered interdependent with the psyche. We have found this connection valid in our own work.

Articular Restrictions (Adhesions and Fixations)

Articular visceral restrictions bring about a loss of mobility and motility because of the inefficient sliding of the organ on its surrounding structures. When they result in a reduction of motility but normal mobility, we call these articular restrictions "adhesions." If both mobility and motility are affected, we call them "fixations." They can be partial or total and are generally the sequelae of infectious pathologies or surgical interventions.

The parietal pleura and parietal peritoneum are partially composed of permeable epithelium. Any infection can cause an abscess (either locally or at a distance from the original site of infection, depending upon its position). These abscesses heal when treated, but this healing is accompanied by adherence to the neighboring tissues. Thus, pleurisy and peritonitis cause fixations which in turn reduce mobility and motility.

Any surgical intervention of the thorax or abdomen unfortunately necessitates the opening of these cavities. The air which enters dries the serous membranes, which tends to promote restrictions. Surgeons cannot prevent this process, although they use their skill to minimize damage to the peritoneum and correctly sew up, layer by layer, the tissues which have been cut.

An adhesion or fixation is thus brought about by a natural or surgical healing process involving the local disruption of normal tissue fibers and their replacement with relatively inelastic granular tissue. It can be conceptualized as a localized drying-out of the attached tissues. The organ will move around the point of fixation, which becomes the new axis of motility and mobility.

Depending on their size and the size of the organ, fixations bring about partial or total articular restrictions. Partial articular restrictions modify the axis of motility — the axis goes through the adhesion. Total articular restrictions completely inhibit motility; the organ becomes inert and loses its rhythm, and its vitality and ability to function properly are affected.

A scar creates a permanent state of mechanical irritation by forcing the tissues to rub against each other. This becomes the focus for a pathological decrease in motion over time because of the large number of movements that an organ makes daily. Attached tissues and organs will find their axes modified, which will cause stretching at the level of the mechanoreceptor system and produce both localized and generalized spasms via a reflex route. The circulation of blood and lymph through the organs will be diminished, causing stasis. The organs will then become prey for multiple dangers — attacks from microorganisms, varicosities, autoimmune processes, stasis of secretory or excretory liquids, and perhaps even malignancy.

Some scars (e.g., from an appendectomy) are visible, but a multitude of others (e.g., sequelae of infectious phenomena) exist without external signs. It is most important that these areas be mobilized. Even when there is a visible scar, this is only the "tip of the iceberg"; think about all the levels that the surgeon has had to sew up which are not even necessarily parallel to the cutaneous surface. When the peritoneum is affected on its sliding surfaces, there are changes in the serosity (amount or clarity) and viscosity (quality or cohesiveness) of serous fluids. These changes are comparable to the effects on synovial fluid of a knee wound which, after several years, ends in arthrosis. Serosity and viscosity of the peritoneal fluid have an important role in the nutritional and immunological status of the articulating structures.

Ligamentous Laxity (Ptoses)

This term refers to a loss of elasticity in the ligament from prolonged overstretching, usually secondary to adhesions. The ligaments, mesentery and omenta are usually only reinforcements of the folds of the peritoneum or the pleura, and are not contractile. Exceptions do occur, particularly in the female urogenital support system, where there are contractile fibers in the ligaments. Another very important exception is the suspensory muscle of the duodenum, or Treitz's muscle in the duodenojejunal region (see page 144).

Visceral ptoses are more often found in a longiline asthenic (flouric, sandlike by homeopathic classification) type than a brevilineal tonic (carbonic, rocklike) type. Hypotonia brings about a laxity of the support structures which decreases the restraints on the viscera. Fat also plays a fairly important role in support phenomena — so much so that one rarely finds renal ptoses in well-built people who have sufficient perirenal fat. Congenital phenomena can be important in the processes of uterine ptosis and retroversion, which is partly attributed to sexual immaturity *(Contamin et al. 1977).*

People suffering from depression are prone to ptosis for at least two systemic reasons. The depression of the central nervous system causes a generalized decrease in tonus, and the weight loss that is so often a part of the depressive syndrome results in a decrease in the supportive fat around the organs. This is an excellent example of how any somatic or psychological event which is sufficient to have an effect on the cortex will also have an effect at a visceral level.

As people grow older, the tissues lose their elasticity, supporting tissues "let go," and range of motion is reduced. The organs direct themselves where gravity takes them, usually downward. Descended bladders, kidneys, uteruses and colons are very common in geriatric populations. The tension of the pulmonary parenchyma is often considered a determining factor in general visceral support, since it decreases with age and thereby tends to promote visceral ptosis *(Kahle et al. 1978; Cruveilhier 1852).*

Multiparity has a strong statistical association with visceral ptosis, but with experience one realizes that it is not the number of childbirths that is important, but rather the manner in which they are carried out. Women who deliver with forceps or suction, whereby the child is "torn out" without regard to the contractions, have their perineum drawn downward at a moment when the tissues, under the influence of hormones, are relaxed and very stretchable. If the obstetrician had a slightly heavy hand, some tissues will never regain their original position and elasticity. Add to this an episiotomy and the associated scar tissues and you will find all the necessary conditions for a variety of abdominopelvic ptoses and dysfunctions. Suction should only be considered in cases of dystocia and is certainly not indispensable. It can have a multitude of adverse effects on the newborn baby as well as on the mother.

Muscular Restrictions (Viscerospasms)

Muscular restrictions or viscerospasms affect almost exclusively the hollow organs. These organs have a double smooth musculature with longitudinal and transverse circular fibers. At rest, these fibers are totally relaxed; in activity, each muscular system contracts alternately to ensure transit. Irritation of a group of fibers (of which there are numerous possible causes) can make them go into spasm, in turn producing a stasis of transit which can be quite significant. The organ no longer fulfills its functions, or carries them out inefficiently. Motility (especially its amplitude) is also reduced. This is a local phenomenon which usually only affects part of the organ. Because the phenomenon is often limited in time, it is similar to a muscular contraction. Gastritis is an example where the irritation causes a reflex restriction which leads to immobility. If this immobility lasts a long time, alkaline and acid residues attack the mucous membrane. This restriction, even when it is not constant, can lead to a duodenal ulcer. With viscerospasm, motility is affected first; mobility is only decreased when the organ's attachments are affected.

The narrower the lumen of the hollow organ, the more important the functional implications of viscerospasm will be (e.g., the secretory canals of the bile duct and pancreas, and the excretory canal of the ureters), particularly if discharge is regulated by a sphincter (e.g., the sphincter of Oddi).

Viscerospasms may have multiple localized or generalized causes. Their appearance is often the first stage of an illness. At the beginning it may be asymptomatic and then bring about functional problems which force the body to use up all its adaptational resources; if these become exhausted, the visceral structures will be affected.

PROBLEMS OF RHYTHM

We have observed that an organ works most efficiently when following a well-defined amplitude and rhythm. The organ traces a motion and then goes back to its original position; after a period of rest, it starts again. A lessening of vitality, besides changing the axes of motion, causes either a lengthening of the rest period or a slowing down or irregularity of the rhythm. This type of problem most often affects visceral motility.

Evaluation

We must remember the importance of the classical physical examination, which is composed of palpation, percussion and auscultation. When examining patients in the context of visceral manipulation, we pay special attention to certain aspects of the classical examination:

- palpation tells us about the tonicity of the walls of the cavities
- percussion tells us about the position and size of certain organs
- auscultation tells us about the circulation of air, blood and secretions such as bile

We have already discussed the effects that muscular activity has on visceral mobility; it is imperative to perform mobility tests of the musculoskeletal system. Please refer to the standard reference works for descriptions of these tests *(Fryette 1980; Mitchell et al. 1979)*. We must stress here our belief that visceral restrictions are the causative lesions much more frequently than are musculoskeletal restrictions.

MOBILITY TESTS

These normally consist of precise movements, to make the organ move directly. The rhythm and direction of the movements are determined by the practitioner. In the case of the liver (see chapter 4), you literally lift it up to appreciate the elasticity of its supporting structures and the extent of its movement. Mobility tests are effective as preparatory techniques to prepare areas for diagnosis. While they require less finesse than the motility or "listening" tests, they are still quite important. They provide information about the elasticity, laxity, spasm or structural injury of muscular and ligamentous structures. Often these tests are able to help us appreciate what is truly happening and clear away assumptions which may be false.

MOTILITY TESTS (LISTENING)

There are numerous methods for testing motility, some using the pulse and others differences in temperature, but for us the most appropriate is "listening," the term used by Rollin Becker. This admirable word conveys the modesty and gentleness which you, the practitioner, should manifest. Listening is the essential modality for the evaluation of the axis and amplitude of motility of any viscera.

Place your hand over the organ to be tested, with a pressure of 20-100g, depending on the depth of the organ. In some cases, the hand can adapt itself to the form of the organ. The hand is totally passive, but there is an extension of the sense of touch used during this examination. Let the hand passively follow what it feels — a slow movement of feeble amplitude which will show itself, stop and then begin again. This is visceral motility! After several cycles, you can try to estimate the frequency, amplitude and direction of the motility. If you have problems feeling the motion and emptying your mind, it may help to focus your attention on the precise anatomical form of the organ to be tested. Conscientiously avoid any preconceived ideas about what is supposed to happen at the beginning; we all possess fertile imaginations, which can mislead us.

Paired organs should be tested together initially; if a problem is found with one, test it individually. Let your hand follow the cyclical motion of the organ around a neutral point. In the chapters that follow, we will describe the position of the hands, the evaluation of amplitude and the axes for specific organs. The frequency of visceral motility, as noted earlier, is 7-8 cycles per minute, slower than that of the craniosacral rhythm and roughly half that of the diaphragmatic respiratory rhythm.

In motility tests, you can check to see if another problem is more primary (e.g., if a dysfunction of kidney motility is secondary to a problem in a nearby organ) by lightly pressing on the other organ. We believe that this inhibits the organ, i.e., decreases its activity in a way that temporarily prevents it from affecting other organs. If the dysfunction you are feeling is secondary, it will markedly improve or even disappear when the structure with the more primary restriction is inhibited.

Manipulations

Before beginning treatment, a diagnosis must be made and the restriction localized and classified (adhesion, fixation, ptosis or viscerospasm). A restriction in one viscera will often cause a restriction in another.

Visceral manipulation is a method of restarting the mobility or motility of an organ utilizing specific, gentle forces. In other words, we provide a stimulus to which the body responds. This concept of restarting the motion implies all the respect which osteopaths should accord to the body — we manipulate to the point where the body can take over in order to achieve self-correction, not to force a correction on the body.

There are three general techniques of visceral manipulation, which either separately or in combination may be the most appropriate in a specific case. They are:

- direct techniques with a short lever arm (for mobility problems)
- indirect techniques with a long lever arm (for mobility problems)
- induction techniques (for motility problems)

Note that we define direct and indirect in terms of the length of the lever being used. Techniques in which the forces are applied locally through a short lever are direct; those in which the forces are applied at a distance through a long lever arm are indirect. This is the common usage of these terms in Europe, and is followed throughout this book. In American osteopathic circles, the definitions of direct and indirect techniques are different, usually revolving around actions relative to a motion barrier. In that system, direct techniques carry the lesioned component through the barrier; with indirect techniques the motion barrier is disengaged.

DIRECT TECHNIQUES

These affect mobility, and are performed with the pads of the fingers of one or both hands, depending on the organ to be treated. First, be sure that nothing you do will cause the patient unnecessary discomfort. Think about the temperature of your fingers; nothing is more disagreeable than cold fingers on the body. Under no circumstances should the contact be with the fingertips of a hand held perpendicularly to the skin, because too much pressure, which is easily applied with the fingertips, is painful. The body reacts to a painful stimulus by contracting to guard against the aggressor; such a reaction will obviously not be conducive to your attempts at treatment. Your fingers should be placed at an oblique angle to the body and the fingerpads used so that the forces applied can be easily controlled.

Direct techniques consist of applying a slight traction first to put the organ, or part of it, under tension, and then to mobilize it while retaining this tension. The organ is mobilized with respect and gentleness, by short back and forth movements performed slowly (usually 10 cycles per minute). These movements encourage the proper direction and amplitude of the motion and improve the elasticity of its supportive structures. As the mobilization proceeds, the tissues become less tense and the tension you apply is progressively relaxed.

Another type of direct technique is called recoil. This also involves putting the organ under tension, but instead of backing off slowly and repeating the pressure rhythmically, you release the tension suddenly. This is usually repeated 3-5 times. This technique focuses the attention of the body on the specific organ and is often used on the organ with the most primary (or least secondary) restriction.

The general principle for treatment of all types of restrictions is to put the organ under tension, using traction, before using specific techniques to mobilize it. For an adhesion or fixation, the organ is placed under tension by a progressive traction applied perpendicularly to the adhesion, with the mobilization performed parallel to it (remember that different planes exist in an adhesion). For a ptosis, the progressive traction takes place in the direction opposed to it (usually upward), with a slight mobilization along the axes of mobility. For a viscerospasm when mobility is disturbed, the direct technique consists of putting the organ under tension and then mobilizing it by encouraging the motion in the direction of greatest mobility.

INDIRECT TECHNIQUES

These affect mobility and involve the use of long lever mobilization to act indirectly on the organ. They are frequently used in combination with the direct techniques described above. For example, in a case of renal ptosis, it is very useful to place the patient in the supine position, bend the legs and, while keeping an upward traction on the kidney, to mobilize the bent legs so that the lumbar column rotates away from the kidney being treated. It is this combination of flexion and lumbar rotation which will indirectly "reposition" the kidney.

Generally, the long lever is used either to mobilize or to increase the effect of tension. It is particularly useful, as we shall see in the following chapters, for organs which cannot be reached by a direct technique (e.g., lungs, mediastinum). When direct and indirect techniques are used together, we speak of combined technique.

INDUCTION TECHNIQUES

The viscera have a pendulum-like motility which is defined for each organ by a direction and an axis. Knowing these axes, you note the amplitude and direction of motion during listening. Physiological motion occurs around a neutral point. With certain restrictions, the motion loses its symmetry, becoming limited in one aspect of visceral motility (inspir or expir) and exaggerated in the other.

To perform induction, you must know the proper and precise directions of motility for each organ. These are described in the chapters that follow. In listening, the hand passively follows the pendulum-like motion of the organ. During induction, the same hand will slightly accentuate or encourage the larger motion (i.e., inspir or expir), which is that in the direction of greater excursion. Continue this process until the induced motion coincides with normal motility of the organ in terms of direction, amplitude and axes. You should almost never attempt to force an increase in excursion of the lesser motion. It is an allopathic concept to wish to force the body. There are exceptional cases when nothing else works and it is necessary to try force. If motility is good (i.e., it is not restricted in either inspir or expir and follows a normal axis), we take the opportunity during induction to slightly accentuate both parts of the cycle. This will increase the vitality of the organ and improve its function. This is why this process is sometimes called facilitation.

Applying induction to an organ that has a significant restriction brings the organ to a point of equilibrium. At this point the motion of the organ may or may not stop temporarily. When it does, this temporary cessation of an inherent motion corresponds to the still point of craniosacral therapy. After a period of time, usually rather short, this still point is followed by a new pendulum-like motion which follows the normal directions of the motility more closely and has a greater amplitude.

There are a few very significant differences between the still point of visceral motility and that of the craniosacral system. While obtaining a still point is almost the *sine qua non* of craniosacral therapy, passage through it is not necessary for a therapeutic effect in visceral induction and will often not occur. When a still point occurs during the course of craniosacral therapy, it is extremely important that the practitioner (in the words of W.G. Sutherland, D.O.) "holds on" until the craniosacral rhythm starts again of its own accord. This may take less than a second or up to half an hour. In visceral induction, when a still point occurs it is best to allow it to continue for 10-20 seconds and then to *gently* restart the motion, first in the direction in which it moved most easily before the induction. If this is done correctly, the visceral motility will return in full force within a minute, be stronger, and follow its proper path.

Sometimes this restarting process will be unsuccessful or the novice practitioner will be unaware of the still point when performing induction. In such cases the still point can last up to several minutes and you may have a problem getting out of it. When this happens, leave that particular organ for a few minutes and attend to another. The motility of the organ which underwent still point will return of its own and be improved. This procedure is in marked contrast to craniosacral therapy, where it is imperative not to remove your hands from the treatment area during the still point. If you utilize craniosacral therapy in your work, it is essential that you understand the differences discussed here. If you are bound by your craniosacral habits while using visceral induction, you will not obtain optimal results.

Induction works better on tissues which are primarily muscular as opposed to ligamentous. We believe this is because muscle fibers are constantly communicating their variations in tone and position to the brain and the rest of the body. Therefore, they "speak louder" than ligamentous tissues, and are easier to "listen to" and treat via induction.

Generally it is best if induction, the treatment of motility problems, is preceded by elimination of the larger restrictions of mobility. It is difficult to release musculoskeletal restrictions using induction, and they can significantly impair motility. The exception to this rule of treatment occurs when the vitality of the viscera is extremely weak, in which case techniques to improve motility should precede those to improve mobility. This is because without a modicum of motility, efforts to improve mobility will have little or no effect.

All of the techniques described in this book should be performed in a gentle, rhythmic manner. Sometimes the motion that we want must be introduced, and sometimes the motion that is already there is accentuated, but always in a rhythmic manner. We believe that this approach is not only successful in accomplishing mobilization, but that it is also the most effective means to reprogram the body by demonstrating to it (both locally and at the CNS level) the proper, healthy, rhythmic motion it should have.

All sessions using visceral manipulation should end with induction of the primary organ involved, as restoration of vitality is the most important effect of this work. For example, a ptosed organ will usually redescend easily even after treatment. The dysfunction of the organ, however, is due primarily to a loss of both mobility and motility, not because it is in a ptosed position. If its proper mobility and motility are restored, its function will improve even if the ptosis remains. At the end of treatment, "listening" should be repeated to check the effectiveness of the treatment.

CONTRAINDICATIONS

A manipulation, even if it resolves a problem locally, should not create other problems. With an infected organ, even if visceral manipulation brings about locally improved function, there is a considerable risk of propagating infection. Where there is risk, it is better to choose nonintervention. This is the law of *primum non nocere.* Therefore, visceral manipulation should not be done in patients suffering from acute infections; the bladder is an exception (see chapter 9).

Foreign bodies constitute a real risk. Be careful about IUDs, calculi and generally anything which could injure the tissues. Manipulating the uterus when it contains an

IUD can cause lesions and hemorrhages — we have seen it happen! We do not say that a kidney with stones should not be manipulated (on the contrary, this is a good indication for manipulation), but you should first be certain that evacuation of the stone is not detrimental, and that migration of the stone cannot cause urinary obstruction and infectious complications. Your evaluation of the patient should also have satisfied you that there is no thrombosis before you begin using manipulations; there are risks from migration of a thrombosis.

A list of possible contraindications would be endless; there will always be a need for your understanding of the osteopathic concept and your common sense. You have everything on your side to keep your patient in good health.

MANNER OF TREATMENT

Law of Precision and Least Force

The more precise the movement, the less force required of the manipulation. Excessive force in osteopathy must be avoided, for it always hides deficiencies in the understanding and skill of the practitioner. One cannot force the body in a direction without putting it under stress. The more force you use, the more you feel your own fingers rather than the patient's response. If your pressure causes pain, the organ being treated will go into spasm and become even more immobilized. If you treat the wrong organ or overtreat, you will further irritate the tissues you are supposedly healing. In this way visceral structures are more sensitive to inappropriate forces than are musculoskeletal structures.

It is essential that you have a sound knowledge of normal human anatomy and physiology; pathology is no more than normal physiology which is speeded up or slowed down. Practice your techniques on healthy people until you become comfortable with them; familiarity with normal motions is a prerequisite to treating abnormal motions.

Rhythm and Amplitude

For each case, you must tune into the patient and feel the rhythm, vitality and resistance of the tissue you are working with. Problems arise from insufficient understanding of the osteopathic concept. One of the big mistakes beginners make is trying to "push" the organ too quickly; the organ cannot adapt to the unnatural speed of change and the treatment is ineffective. Your role is to restart the organ; the body will take over from there. If, after treatment, the organ goes through about 10 normal cycles, you may consider the treatment to have been appropriate.

The amplitude of motility is never very great; we will discuss this in greater detail with the description of each organ below. As there is more amplitude in visceral mobility, direct or indirect manipulations to improve mobility require a bit less finesse than do those which affect motility. It is important to remember that all manipulation must be done with care and understanding.

Number and Frequency of Treatments

It serves no purpose to treat a patient too often. You cannot replace the vitality which the patient is lacking; rather, you must stimulate the patient's forces of autocorrection. An interval of three weeks between sessions has given us the best results over the last fourteen years, with thousands of patients. We usually see a patient three times and ask him to return six months or one year after the last treatment.

The timing of the appointment may be important. If you are working on the urogenital system, it is not advisable to treat a female patient just before her period. The congestive premenstrual conditions restrict the organs and decrease the effectiveness of treatment. Similarly, manipulation of the stomach is easier to carry out if the stomach is empty. Seasonal considerations can be important. For example, it seems that hyperactive patients are particularly sensitive around the equinoxes and preventive treatment just before these times will be very helpful.

Length of Treatment Sessions

This is difficult to quantify, because each individual and each organ is different, and practitioners have different levels of skill. In general, restoration of mobility to an immobile organ is a criterion of success, and 10-15 cycles of direct or indirect movements should achieve this. *If there is no improvement after 15 cycles of treatment, either you are performing the technique incorrectly or the problem is someplace else.* If you persist in continuing the technique, you will only irritate the tissues.

It is easier to improve mobility than motility; treatment of the latter usually requires more skill. The more precise your manipulation, the less intervention will be necessary. This refers to *both* the amount of treatment at each session and the number of treatment sessions to achieve a given result. We call this the goal of treatment *a minima.*

You can overtreat a patient (i.e., attempt to make more changes than the body can handle) during a single treatment session or by spacing the sessions too close together. When this happens, the tissues will "buzz" in a peculiar and unmistakable manner. This sensation resembles the sticky or "tacky" feeling that results from overtreatment of the craniosacral system, although the two phenomena are distinct.

Effects

It is not possible to produce an isolated effect upon the body, as it is an integrated unit. Simple manual contact will have prolonged effects; one can never equate visceral manipulations with a simple emptying or lifting of an organ. If you are in tune with the inherent motions of the body, all processes, both localized and systemic, can be affected.

At the most basic level there are unicellular axon reflexes, comparable to the muscular stretch reflex, which will then reach other segments. These reflexes will have a role in the tonus of the organ and the sedation of spasms. The stimuli will excite the whole spinal cord, then the different cerebral centers, using the reticular activating system as a relay. It is now generally recognized that acupuncture excites the production of certain neurotransmitters in the brain (e.g., endorphins, serotonin, dioxyphenylalanine), which, in turn, stimulate other centers (e.g., hypothalamus, pituitary, thyroid, adrenals) and thereby cause secretion of hormones (e.g., FSH, TSH, adrenalin). We believe this to also be true of other manual stimuli, including those used in visceral manipulation.

It is the present scientific opinion that serotonin is synthesized in the tissues where it is found, and not in a specific organ from which it is transported to other organs via the bloodstream. In mammals, most serotonin is found in the gastrointestinal tract, but it is also an important central neurotransmitter. Increased synthesis of serotonin leads to stimulation of cerebral activity and smooth muscles of the blood vessels, digestive tract and respiratory tract. We believe that visceral manipulation increases tissue metabolism which in turn stimulates general metabolism via increased serotonin production. This has been difficult to prove because serotonin levels fluctuate whenever a subject is touched. This could explain certain reactions to visceral manipulation which seem disproportionate to the intensity of the treatment. Important examples are headaches (including migraines), irregular periods and galactorrhea. With experience you will be able to add to this list.

Evaluation of the effects of treatment can be problematic. Often, after the first treatment session, there is a hyperreaction caused by the body's difficulty in adapting to an unknown stimulus. It is up to you to make the most of this first session and, above all, to warn your patients of this possibility. We are aware of the placebo effect and cannot deny its effects in our overly sensitive or suggestible patients. We usually base our conclusions on results from treatment of independently-minded patients who demand fast and efficient treatment and who have little interest in staying inactive. It is more difficult to evaluate treatment of female patients — the physiological variations, due to their hormonal rhythms, are a real challenge to practitioners. For example, the effects of liver manipulations are very different premenstrually than postmenstrually.

In general, we believe that visceral manipulation affects:

- mobility and motility
- circulation of fluids
- sphincter and muscular spasms
- hormonal and chemical production
- immunity (both localized and systemic)
- the psyche

Adjunctive Considerations

In the following chapters, after describing techniques for manipulating each organ, we will mention some of the principal osseous restrictions associated with visceral pathologies. Do not lose sight of the global concept of a total lesion. These osseous restrictions are present, and it is up to you to decide if they need adjusting or not. Manipulating a vertebra which is not restricted, just because it is related to an organ that has a problem is, for us, a betrayal of the osteopathic concept. Only that which has a restriction of motion should be manipulated.

Some practitioners use diaphragmatic respiration as an aid when manipulating an organ. We have done this in the past but now prefer to manipulate during normal respiration as in this situation the natural resistance of the viscera is present. Patients asked to breathe deeply must contract the abdominal muscles to do so. When working on such cases you have to increase the force you use, which makes effective treatment more difficult.

Finally, we will include some recommendations for adjunctive measures. Exercise has often been helpful in reinforcing the abdominal musculature, which will aid in restoring proper functioning of the viscera. We will include advice about diet, and mention indications and contraindications of certain physical activities. For example, encourage patients with ptosis problems to use the reverse Trendelenburg position as much as possible.

Clinical Example

In this section, we shall describe a typical example which demonstrates our manner of reasoning and treatment.

Mr. X, 42 years old, presented himself complaining of a chronic right cervicobrachial neuralgia, which he had for ten years. This caused a constant discomfort interspersed by sudden exacerbations. The pain was particularly debilitating and demoralizing because he was a skilled joiner.

There was no history of trauma. Mr. X had never been ill enough to seek treatment (except for pain); he had a vague memory of several days of fever, without apparent cause, during his military service.

The mobility tests showed an articular costovertebral restriction of T1 and R1 on the right, a restriction of the right sternoclavicular articulation, and a musculoligamentous restriction at T7/8.

The motility tests showed a visceral articular restriction (fixation) of the right upper lobe of the lung (the axis of motility was skewed), a right deviation of the mediastinum, and perturbed motility of the stomach.

Physical examination showed a decrease of the radial pulse when the right arm was abducted and externally rotated. There was considerable air in the stomach. Chest x-ray showed a scarred pleural adhesion on the lateral edge of the upper right lobe, which the radiologist attributed to a probable sequela of an asymptomatic pleurisy.

The pleural adhesion became the center around which the motion of motility occurred for the superior right lobe; the forces of intrathoracic pressure became modified; the mediastinum was drawn toward the restricted lobe; and the suspensory ligament of the right pleural dome became fibrosed. The cervicopleural fibers were retracted and this retraction fixed T1, R1 and the clavicle which, in turn, disturbed all mechanics of cervical motion. Ligamentous tension compressed the intervertebral foramen and thus diminished the phenomena of fluid exchange.

The clavicle and R1, fixed one upon the other, reduced the thoracic outlet, provoking vasomotor problems of the right subclavian artery by reflex and/or direct compression.

The mobility of the superior right lobe had changed. The respiratory movements (over 20,000 daily) increased the mechanical disorder. The mediastinum, drawn toward the scarred lesion, exerted traction on the esophagus, changing it so much that the sphincter of the cardia became dysfunctional, facilitating passage of air into the stomach. The esophagus become irritated and went into spasm, causing symptoms similar to those of a hiatus hernia.

From this example, which is not unusual, one can draw several conclusions. Small causes can bring about important effects if the former are multiplied thousands or millions of times, even though many people seem to have difficulty comprehending this idea. A slight imbalance of the lower limbs can very quickly ruin even the best pair of shoes. The symptom may be only a manifestation of a far-off disorder; this requires the practitioner to search throughout the body for the cause. The patient is determined to bring the practitioner's eye to where the symptom is, because that is where the suffering is. In our example, the patient was most surprised, even nonplussed, when we said we wished to see a chest x-ray. The relationship between sore fingers and the lungs was difficult to comprehend — it even took us a while to figure it out!

The patient is well now, although the x-ray still shows a pleural adhesion. Obviously, we were not able to break it up and the cervicopleural tension still exists, although much lessened. But motility has returned to normal. This underlines its importance in energetic phenomena. It is of little importance if the organ is in an abnormal position as long as it has recovered its motility.

Summary

The osteopath is a mechanic in the noblest sense of the word — really a micromechanic. We all have two hands but who among us really knows how to use them? No one argues with the wine taster who, by using his palette, can tell us the characteristics of a wine — its region, its vineyard or even its vintage. The education of touch can go at least as far. Through the use of the concepts and techniques discussed in this book, you will be able to improve not only your palpatory and treatment skills, but also broaden your therapeutic horizons.

Chapter Two: The Thoracic Cavity

Table of Contents

CHAPTER TWO

The Thoracic Cavity

It may surprise you that we discuss the lungs and the mediastinum in a book concerning visceral manipulation, since for all practical purposes only the pleural dome is directly palpable. However, by means of stretchings, passive movements and induction, it is possible to functionally palpate and affect the pleura and the mediastinum. Nevertheless, we wish to emphasize that the lungs are comparatively difficult to treat. Beginners at visceral manipulation are advised to start with the abdominal organs. We have placed the discussion of the thoracic cavity here to maintain a top to bottom sequence of presentation.

Our research on visceral manipulation began with the thoracic cavity. Among our patients, many suffered from pulmonary tuberculosis and had been treated by therapeutic pneumothorax. At the time, this technique was used as the principal remedy in accelerating the healing process of the pulmonary parenchyma in patients with tuberculosis. The treatment obviously had its side-effects, including serious scarring, interstitial fibrosis and pleural adhesions. These sequelae bring about changes in the mechanism of pulmonary ventilation by abducting the mediastinum and, frequently, the thoracic vertebral column as well. This illustrates the impact that a repeated movement can have on a pathologically modified axis.

Clinical experience demonstrated to us that a significant number of scolioses are caused by similar problems of the lungs and/or mediastinum. These observations led us to the following hypothesis: all pathology, whether mild or severe, of a structure attached to the thoracic cavity is capable of modifying the mobility and motility of that structure, principally by changing the axes of its motion.

Anatomy

The thoracic cavity is bounded anteriorly by the sternum, laterally by the rib cage and posteriorly by the vertebral column. It is limited superiorly and inferiorly by two diaphragms:

- the upper (superior) diaphragm consists of a musculoaponeurotic system and the suspensory ligament of the pleural dome
- the lower (inferior) diaphragm consists of the diaphragm muscle itself

The thoracic cavity is divided into two lung cavities by the mediastinum. It has two types of relationships with the musculoskeletal structures of the thoracic cage and the two diaphragms (defined below).

RELATIONSHIPS

From back to front, the thoracic cavity is connected with the spinal column, costotransversal joints, costovertebral joints, ribs, costochondral and sternochondral joints and the sternum itself. This list may appear overly fastidious and therefore of no use. In fact, the opposite is true because all the structures of the thoracic cage, whether articular, ligamentous or muscular, affect the mobility and motility of the enclosed viscera. It is not only the osteoarticular structures which affect visceral function.

The inferior and superior borders of the thoracic cavity are enclosed in different ways. Inferiorly, the diaphragm muscle encloses the thoracic cage completely. The only communicating points are the three principal orifices which create passages for the aorta, esophagus and inferior vena cava, along with some secondary orifices. The aorta crosses the diaphragm just anterior and slightly to the left of the vertebral body of T12. This is a fibrous orifice, in the form of a stringy arcade which forms the crura of the diaphragm. The inferior vena cava crosses the phrenic center at the level of T8 and clings closely to these stringy fibers. The esophagus crosses the diaphragm at its fleshy part, slightly anterior and to the left of the aorta, facing T10. The esophagus is joined to the diaphragm by dense connective tissue and muscular fibers. The vagus nerve follows the course of the esophagus.

In effect, the vascular orifices in the diaphragm are sinewy. Anatomical knowledge enables you to hypothesize as to the location and type of pathology. For example, the action of the diaphragm has little effect upon the aorta but, thanks to its sinewy characteristics, acts as a kind of pump for the vena cava, assisting return circulation, normally without any restrictive effects. However, any hypertonia of the muscular fibers of the diaphragm would have an effect on gastroesophageal physiology because of the connective tissue relationship between the esophagus and diaphragm.

The superior diaphragm is much smaller and actually consists of two lateral diaphragms made up of musculoligamentous fibers separated by the mediastinum, which opens onto the cervical region. This area is sometimes referred to as the thoracic inlet (or, less commonly, thoracic outlet) and is a weak point involved in numerous syndromes affecting the upper limbs. All the soft tissues joining onto the first ribs, clavicles and T1 help make up these diaphragms. The most important element is the suspensory ligament of the pleural dome (pages 38-39).

VISCERAL ARTICULATIONS

The thoracic cavity is divided in half by the mediastinum, which contains the heart, esophagus, trachea and vagus nerve. The sliding surfaces between the thoracic viscera and their surrounding structures consist of two serous membrane systems: the pleura for the lungs and the pericardium for the heart.

Pleura

The pleura, the serous membrane system of the lungs, consists of two layers: the visceral layer (covering the lungs) and the parietal layer (lining the walls of the two lung cavities). As in all serous membranes, there is a small amount of lubricating serous fluid between the two layers.

The lungs permanently occupy all the space available for them. During respiration, they do not expand and contract in a uniform manner because some areas move more than others. The lungs are divided into lobes which are separated by deep fissures, the latter being lined by visceral pleura. This lining enables the lobes to slide one upon the other, an important factor in mobility.

The visceral pleura, which is only lightly attached to the lungs, is made up of mesothelium, a layer of collagenous and elastic fibers and a subpleural layer containing the lymph vessels.

In contrast, the parietal pleura is tightly attached to the adjacent tissues. The structure of the fibrous layer varies from area to area, with collagenous fibers being predominant in the rib cage and pericardial areas and elastic fibers predominating around the diaphragm. Above the diaphragm and the intercostal areas is a succession of lymphatic vessels. The parietal pleura is permeable and able to absorb air and liquid from the pleural spaces.

The pleura is usually divided into four regions:

- costal pleura
- mediastinal pleura
- diaphragmatic pleura
- pleural dome

The costal pleura is firmly attached to the ribs, costovertebral groove, sternum and sternocostalis muscle. External pressure is such that the first rib, and often others, makes an impression on the lungs. The mediastinal pleura unites the organs of the mediastinum from the sternum to the vertebral column. With the costal pleura, it forms a costomediastinal cul-de-sac. The parietal pleura covering the mediastinum is broken up by the roots of the lung. The line of reflexion of the two layers of the mediastinal pleura meets the diaphragm, forming the lung ligament (page 39). The diaphragmatic pleura is closely attached to the diaphragm and covers it entirely. The diaphragmatic and costal pleura form a cul-de-sac, which varies depending upon respiration and is called the costodiaphragmatic recess. The pleural dome caps the top of the lungs. The parietal pleura is found here, reinforced by fibromuscular fasciculi (collectively forming the suspensory ligament) which attach the pleural dome to the structures of the thoracic inlet.

Pericardium

The pericardium is composed of a serous membrane and a fibrous pouch. It surrounds the powerful heart muscle, making it difficult to apply the listening system to the heart or to treat it. If there are scars from pericarditis, myocardial infarction, or other processes, you should be able to listen to the scarred fibers, but the strength of the cardiac rhythm makes listening very difficult. In this way it is similar to (though much stronger than) the respiratory motion of the lungs. However, because the pericardial pouch can be listened to, it is interesting to examine briefly the envelopes of the heart and their relationships with the lungs.

The pericardial serous membrane is made up of two layers, a visceral layer and a parietal layer, which delineate a virtual space called the pericardial cavity. The linings of this cavity are moistened by a serous fluid which facilitates the ability of one layer to slide upon another. The parietal layer is directly attached to the fibrous pouch, while the visceral layer covers the heart from the bottom to the top of the aortic hilum, which it sheathes, and there joins with the parietal layer.

The fibrous pouch, a very thick membrane, is actually a doubling of the parietal layer. It is made up of interlacing collagenous fibers and its role is to prevent any noticeable dilatation of the heart. The heart is attached to the skeleton by ligaments which insert themselves into this pouch.

Lungs

The lungs are attached to their surrounding structures by a suction system and the suspensory, lung and interpleural ligaments.

The suction system is created by the negative pressure within the thoracic cavity, which forces the lung to be always flattened against the lining. The suction is localized at the periphery of the lungs and makes possible thoracic expansion.

The suspensory ligament attaches the pleural dome to the skeleton *(Illustration 2-1)*. It consists of muscular fibers of the scalenus minimus (sometimes mixed with fibers

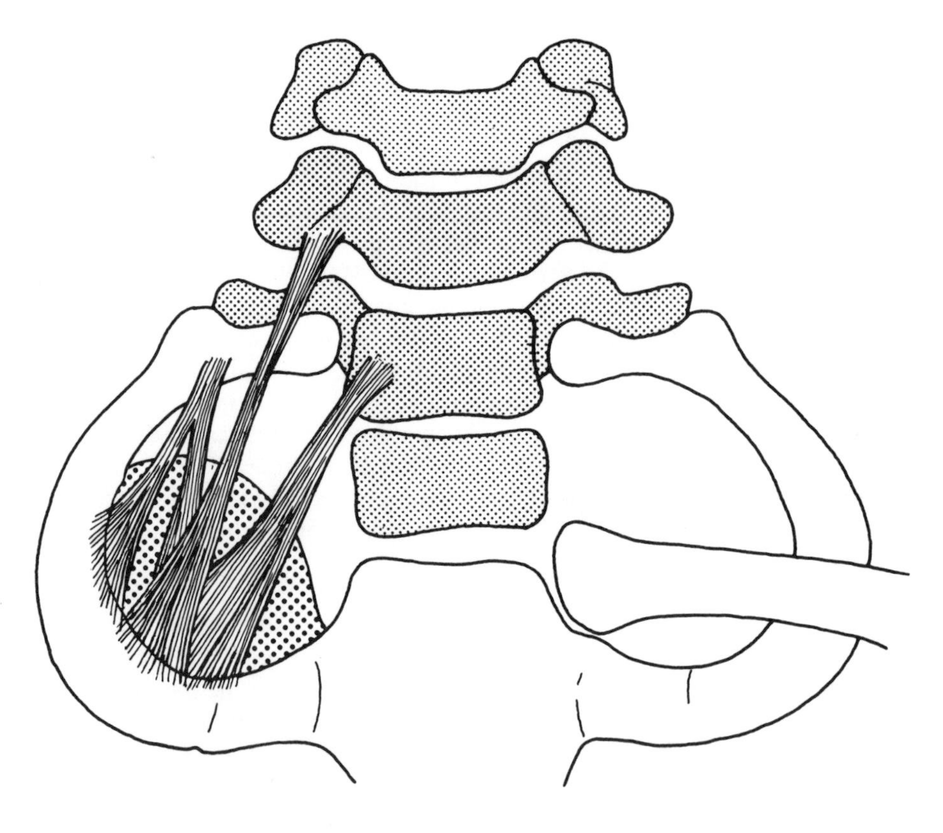

Illustration 2-1
The Suspensory Ligament of the Pleural Dome

of the anterior and medial scalene muscles), plus the fibrous fasciculi (page 37). This ligament is not directly inserted into the parietal pleura, but rather into the intrathoracic fascia. This fascia forms a "connective tissue dome" at the level of the top of the lungs, where it and the elements of the ligament form a partition. This partition, which is anatomically independent of the parietal pleura, is firmly attached to the nearby skeleton and is called the cervicothoracic fibrous septum. In the physiology of movement, this septum is the link between the superior lobe of the lung and the cervicothoracic junction.

The lung ligament is usually said to be formed from reflected folds of pleura under the pulmonary hilum. In fact, the fold does not stop at the pulmonary hilum, but continues as far as the diaphragm. Overall, the line of reflexion has the form of a tennis racquet, with the web-like part surrounding the pulmonary hilum in the front, behind and above, while the handle is represented by the lung ligament, which is connected to the thorax like a mesentery. Both strips of this "mesentery" are joined together *(Illustration 2-2)*. The lung ligament is linked to the esophagus by means of surrounding fascia.

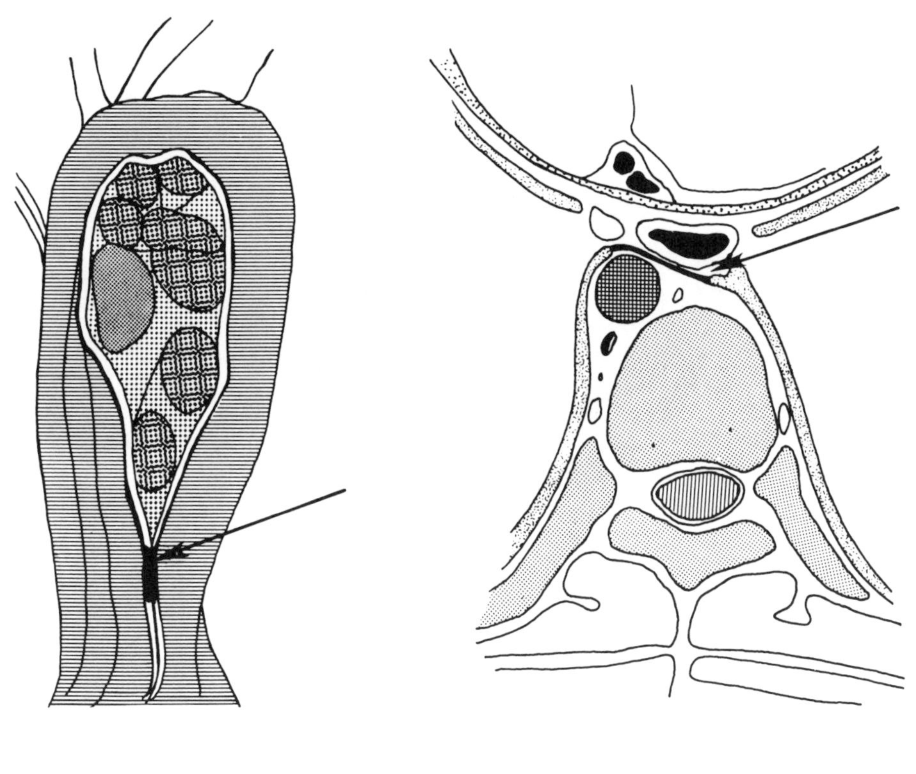

Illustration 2-2
The Lung Ligament

Illustration 2-3
The Interpleural Ligament

Another link between the two lungs is the interpleural ligament, which is formed by the joining of the two interazygos cul-de-sacs *(Illustration 2-3)*.

Heart

The heart is attached to surrounding structures by a suction system and a ligamentous system. The suction system is the same as that of the lungs. There is a parietal membrane lining the deep face of the fibrous pouch and a visceral membrane covering the heart. The fibrous pouch is therefore lined with the parietal pericardium on the inside and the mediastinal pleura on the outside. Thus, there is a double system of serous membranes around the heart.

The heart is stabilized as follows *(Illustration 2-4):*

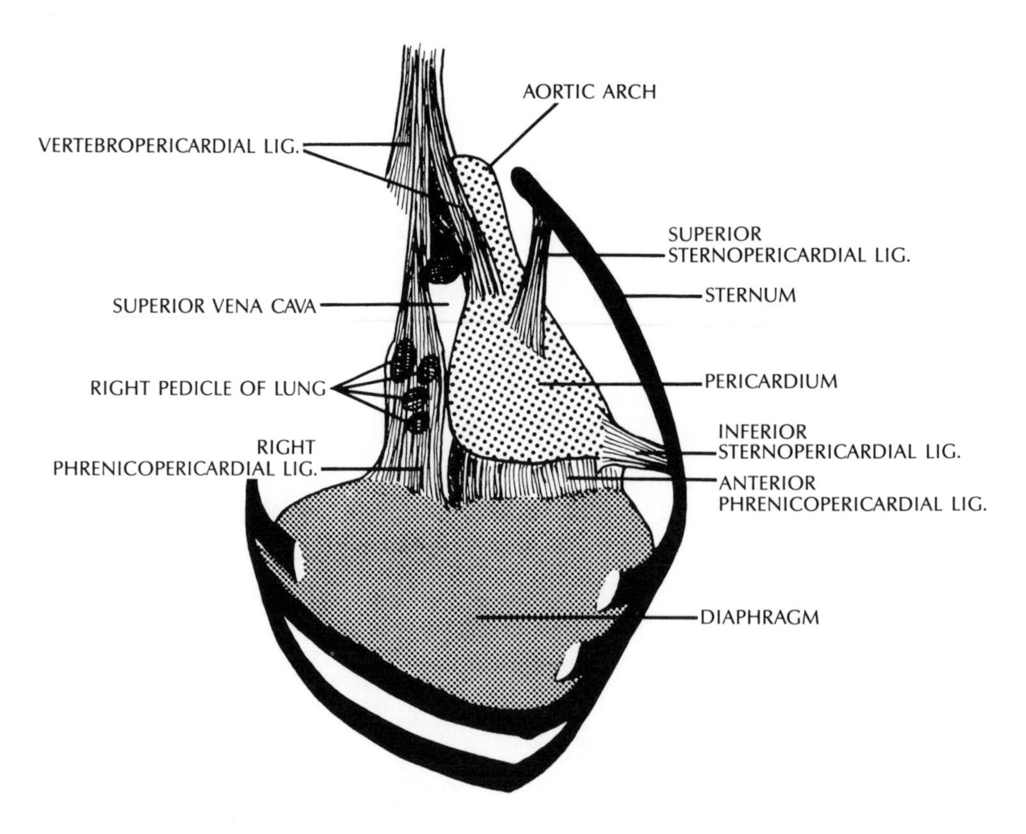

Illustration 2-4
The Pericardial Ligaments

- anterosuperiorly by the superior sternopericardial ligament
- posterosuperiorly by the vertebropericardial ligament
- anteroinferiorly by the inferior sternopericardial ligament
- posteroinferiorly by the left and right phrenicopericardial ligaments
- inferiorly by the anterior phrenicopericardial ligament

Note that the heart has no lateral osseous attachments and that this role is performed by the lungs and pleura. The lungs, because of their tendency to expand, create a medial pressure which holds the heart in place.

TOPOGRAPHICAL ANATOMY

For evaluation and treatment, it is necessary to know where to place one's hands. A knowledge of topographical anatomy gives you a window by which to gaze into the interior of the body. If the line on the left between the border of the sternum and the nipple is divided into three parts, the lungs occupy the lateral third, the costodiaphragmatic sinus the middle third and the heart the medial third. This simplification may startle anatomists, but is largely sufficient for the application of our treatments.

Bronchi

The essential reference mark is that of the tracheal bifurcation. This is at the level of T4/5 posteriorly and the sternal angle or manubrium anteriorly. The most common mistake is to think that it should be more inferior. The bronchi are obliquely directed inferiorly, internally and posteriorly. The angle is more oblique on the right than on the left. The traction of the right lung is stronger, so much so that the tracheal bifurcation is slightly to the right of the median line. The right bronchus is shorter than the left, but larger. The left bronchus is concave superiorly and laterally.

Lungs

The important reference marks are the inferior and superior limits of the lung, costodiaphragmatic antrum, posterior left mediastinal antrum, interlobal fissures and hilum.

The pleural dome is a few centimeters higher than the upper limit of the thoracic cage, which is formed by R1 and the C7/T1 articulation. The pleural dome is the only palpable part of the pleuropulmonary area.

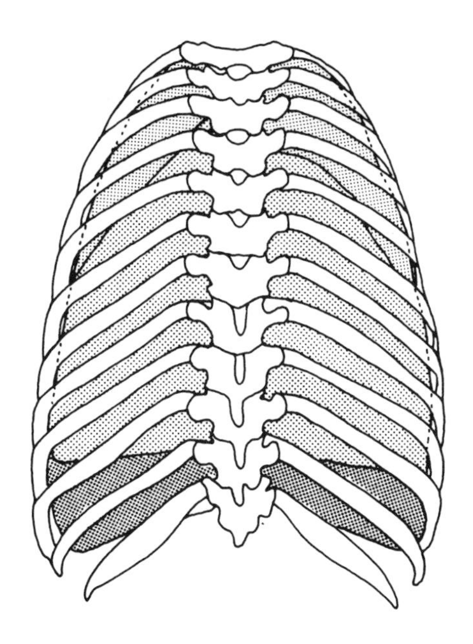

Illustration 2-5
Posterior Pulmonary Reference Marks

The inferior border of the lung, in a position of medium inhalation, lies behind a horizontal line passing through the upper part of T11. Anteriorly, for the right lung, this inferior border is schematized by a line coming from the sixth costochondral joint, via R6 medially, obliquely inferolaterally to the junction of the midaxillary line and R8. Posteriorly, the oblique fissures begin at T4 and go obliquely downward to the meeting point of R6 and the midclavicular line *(Illustration 2-5).*

Anteriorly and on the right, the lungs are delineated inferiorly by a line that is slightly concave superiorly and goes from the lateral edge of the lower extremity of the sternum to the intersection of the axillary line with R8 (see above). The costodiaphragmatic sinus begins in the same place but is oblique and goes down to the intersection of the axillary line and R9. The oblique fissure appears laterally and follows R6. The horizontal fissure begins laterally at the intersection on the oblique fissure of the midclavicular line and R6, and then follows the 4th costal cartilage medially *(Illustration 2-6).*

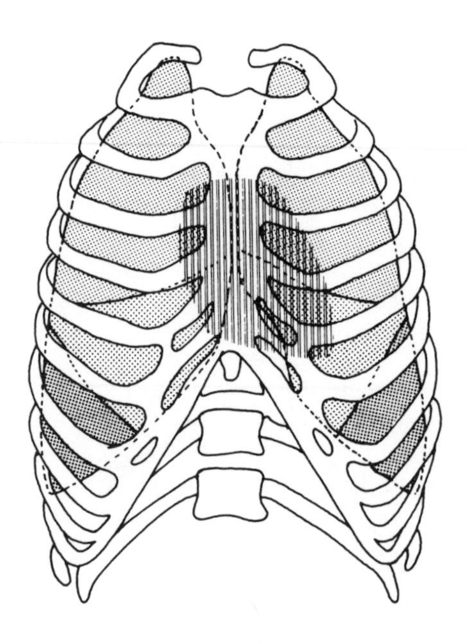

Illustration 2-6
Anterior Cardio-Pulmonary Reference Marks

Anteriorly and on the left, the lung and costodiaphragmatic sinus begin at the level of the sternum and 4th left intercostal space. The heart forms a concave indentation medially and posteriorly. The edge of the lung and the sinus then descend vertically to the 6th and 7th costal cartilage respectively.

The two halves of the costodiaphragmatic antrum begin at the lower extremities of the costomediastinal antrums and then separate laterally, crossing R10 on the axillary line. From this point they go posteriorly, medially and finally superiorly to the 12th costovertebral joint.

Heart

In the normal thorax, the cardiac area is in the form of a quadrilateral. The two superior angles are on each side of the sternum, in the 2nd intercostal space, approximately 2cm to the left and right of the sternum. The right inferior angle corresponds with the sternal extremity of the 6th right intercostal space. The left inferior angle is usually found in the 5th left intercostal space, just inferomedial to the left nipple.

Physiologic Motion

MOBILITY

Lungs

The lungs are in perpetual motion, from pulmonary respiration (the most noticeable movement), skeletal muscle movements and motility.

We have already discussed the pulmonary suction system (page 38). This system keeps the lungs adherent to the pleura but also able to slide upon it. In all movement, the lungs and thorax are interdependent; each lung follows its hemithorax. Obviously, there is no bulk displacement of the lungs and expansion is made in the same directions and axes as the movements of the thorax.

Let us examine what happens during forced inhalation, which is only an exaggeration of normal inhalation. Each hemithorax increases in volume, followed by the pleura and lung. This increase is made possible by the mobilization of the supple structures of the hemithorax. The diaphragm (and diaphragmatic pleura) descends, the rib cage undergoes an anterior and lateral expansion and the costal pleura follow the action of the rib cage. Therefore, the expansion of the hemithorax, and thus the lung, takes place because of the lowering of the diaphragm and costal expansion. The pleuromediastinal wall is fixed.

Because the superior diaphragm of the thorax is essentially made up of stringy structures, the pleural dome is fixed. These fixed points are necessary so that a structure may be stretched. The lungs must be submitted to traction which depends on a direction associated with an exerted tension or force, which in turn depends upon the same axis but in the opposite direction.

The lung is submitted to a force F on the costal pleura, but also to a tension T on the mediastinal pleura, which prevents the entire lung from moving laterally *(Illustration 2-7)*. This tension, which balances the lateral costal expansion, is created via the lung ligament. The tension which balances the downward expansion caused by the diaphragm muscle is created by the suspensory ligament. The movement of the thorax is the sum of the movements of each costovertebral unit (a thoracic vertebra and its pair of ribs).

During inhalation, each lung executes a rotation around a set of axes going through the costovertebral and costotransversal articulations. This axis varies from a nearly frontal plane for the upper ribs to a nearly sagittal plane for the lower ribs. The direction of the axis is directly related to the orientation of the transverse process, which varies in the same way.

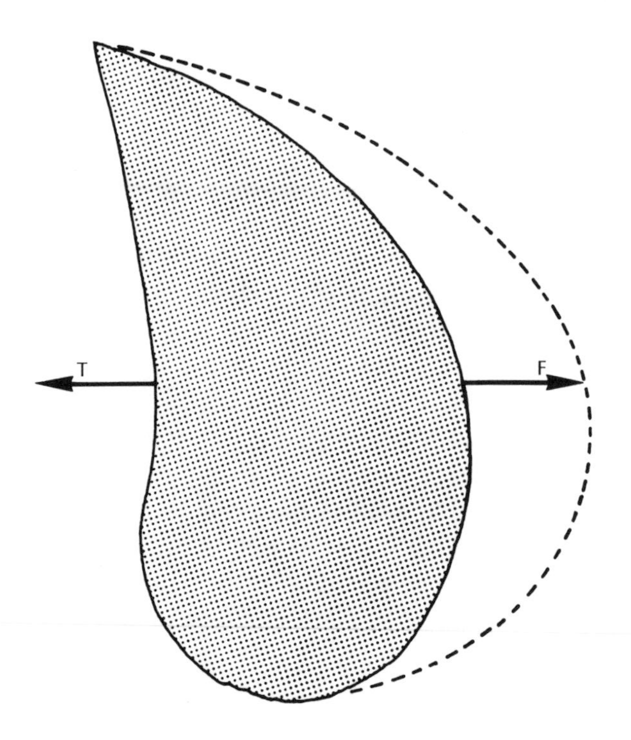

Illustration 2-7
Forces and Tensions on the Pleura During Inhalation

The upper ribs undergo a "pump handle" movement which causes the anterior rib cage and sternum to move anterosuperiorly. The lower ribs undergo a "bucket handle" movement which causes a lateral elevation of the lower ribs.

Another rib movement, which is particularly noticeable during forced inhalation but also exists in a latent state during normal inhalation, is a horizontal rotational movement around a vertical axis. For most costovertebral units, this axis passes through the center of the imaginary circle in which the posterior arc of each rib is inscribed.

In fact, if schematically the portion of the oval which each rib makes is extended tangentially, we obtain an ovoid form with two centers *(Illustration 2-8)*. Each hemithorax has a common anterior center and an individual posterior center. During forced inhalation, each rib rotates externally around its posterior center.

Most of these rib movements increase all the diameters of each hemithorax. The lungs, being elastic, will increase in volume during this external rotation and expansion. In fact, the extension of the arcs of the ribs represents the mediastinal pleura and hemithorax *(Illustration 2-8)*.

The lung, which is fixed to the mediastinum, will be laterally stretched around the posterior center of the hemithorax *(Illustration 2-9)*. On a pulmonary level, this center is made up of the segmentary apical bronchia for the upper lobe and by the bronchial tree for the rest of the lung. The position of the bronchial tree in the lung is logical, as it prevents the whole of the lung from being pulled in all directions during respiration.

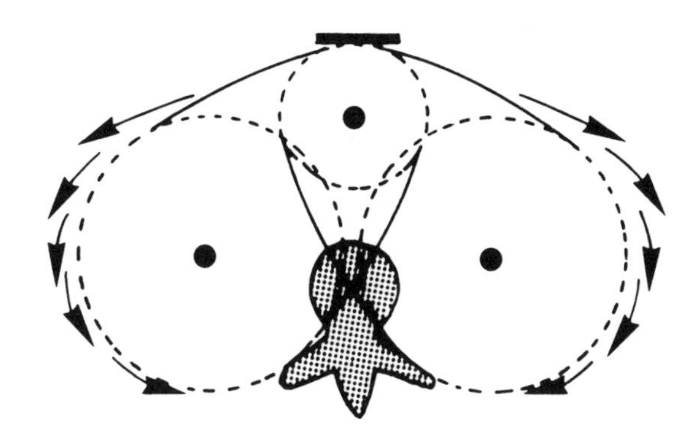

Illustration 2-8
Horizontal Rib Rotation During Inhalation

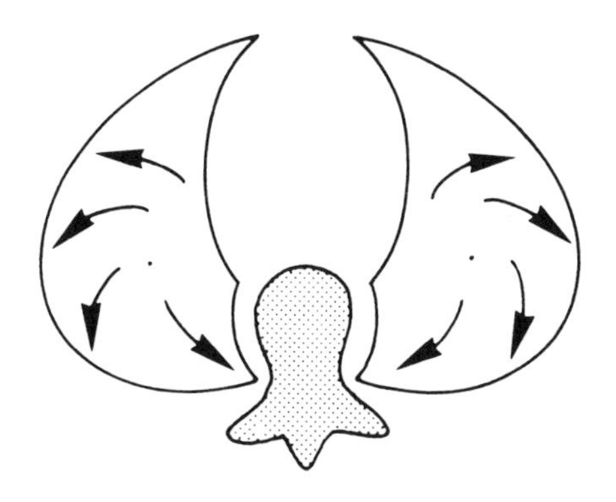

Illustration 2-9
Pulmonary Expansion During Inhalation

All these rib movements are synchronous. The mediastinal pleura is fixed and the stretching of the pulmonary parenchyma follows an external rotational movement. Pulmonary expansion is maximal toward the front for the upper lobes (pump handle) and toward the sides for the lower lobes (bucket handle). During mobility, the middle lobe on the right moves in the same manner as the upper lobe.

In the lower lobes, the bronchial trees are oriented obliquely laterally. External rotation of the lung during inhalation occurs in a plane perpendicular to this axis. Note that the left bronchus comes off the trachea less obliquely than the right *(Illustration 2-10).* The change in the axis of the lung is not restricting, because the resulting torsion is counteracted by the elasticity of the parenchyma and the sliding of the pleura and fissures.

In conclusion, the mobility of the lung during inhalation is an external rotation of the parenchyma on a vertical axis for the upper lobes and on an obliquely inferolaterally directed axis for the lower lobes. Pulmonary expansion is possible owing to the

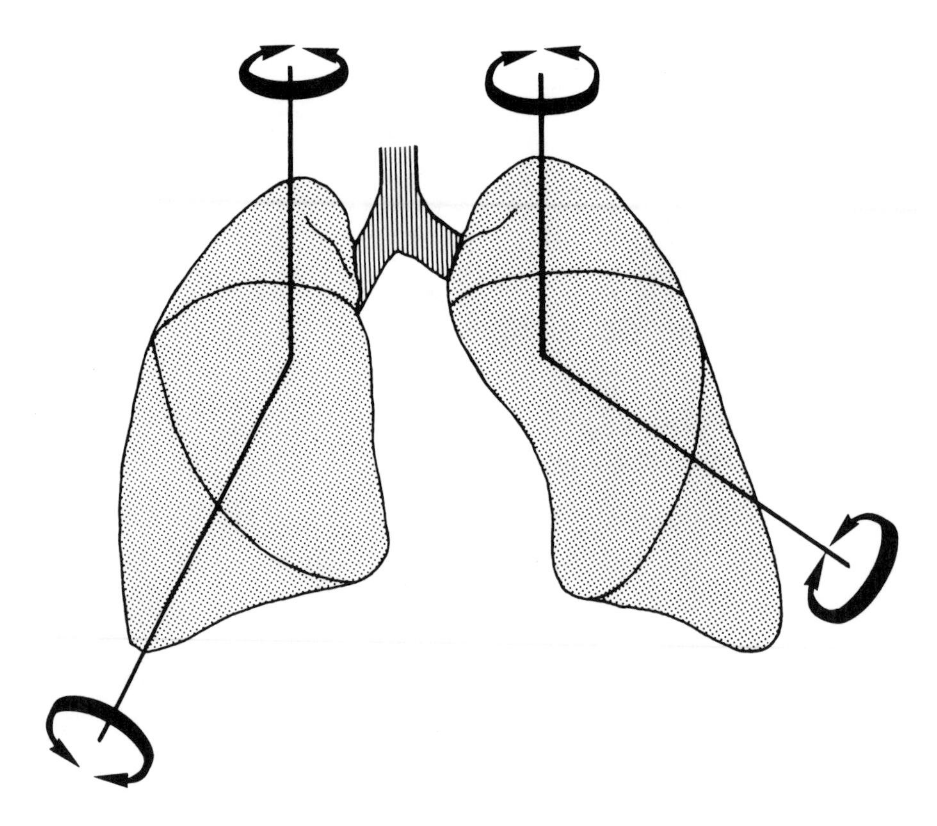

Illustration 2-10
Lung Mobility

tension of the lung ligament, the left bronchus (which fixes the visceral pleura to the mediastinum) and the suspensory ligament (which fixes this dome at the top).

Mediastinum

The mediastinum is made up of the heart and a set of tubes which transport air, blood, water and food. It is bounded by two sagittal mediastinal pleura laterally, the sternum anteriorly and the spine posteriorly. In this section, we will first describe cardiac mobility and then the mobility of the remainder of the mediastinum.

Heart movements are the most frequent (100,000 movements per day) of all autonomic movements. Apart from the vibrations, which spread to the adjacent viscera and are carried to all structures by arterial pulsations, we have detected no repercussions of this pump upon the thoracic viscera. The heart itself performs a powerful torsion (twisting) movement, which is put through a system of shock absorbers. From deep to superficial, this system consists of:

- the double-layered pericardium (which permits sliding)
- the pericardial stringy pouch (which prevents excessive dilatation of the heart)
- the mediastinal pleura
- lateral pulmonary pressure

During inhalation, the ligaments of the lung and bronchi exert an isometric tension on the lungs so that they do not move laterally as a block. With thoracic and pulmonary expansion (force F1), the right lung ligament and right bronchus react with an isometric tension T1, affecting the visceral pleura. Similar forces F2 and T2 are applied to the left lung. Forces T1 and T2, being of equal force and in opposite directions, cancel each other out. The interpleural ligament, which is the union of the cul-de-sacs of both the aorta and esophagus, connects the right and left mediastinal parietal pleura. Forces F1 and F2 balance each other across the mediastinum and should remain in equilibrium *(Illustrations 2-11-A and 2-11-B).*

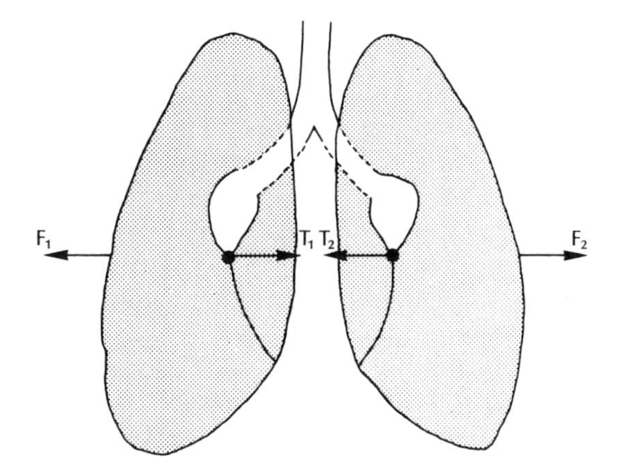

Illustration 2-11-A
Traction on the Mediastinum

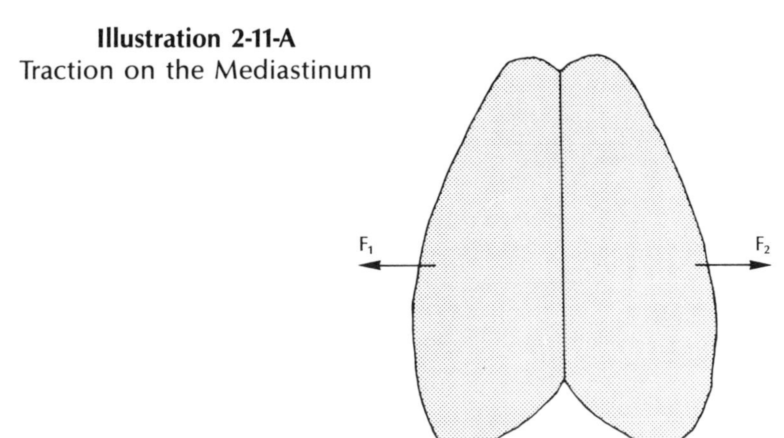

Illustration 2-11-B
Schematic Representation of Mediastinal Tension

During inhalation, the diaphragm is lowered to enable the phrenic center to influence the viscera. By means of a change of pressure, the diaphragm will raise the ribs laterally. These are the basic effects of respiration on the thorax that we were taught in school.

In fact, the situation is a little more complex. The phrenic center certainly does have an influence on the visceral mass, but the vertical tension of the mediastinum intervenes. For purposes of visceral manipulation, *the diaphragm is suspended from the mediastinum.* Do you think that the phrenicopericardial ligaments exist in order to stabilize the heart and to keep it in place? Certainly not. They quickly fix the phrenic center when the diaphragm descends. The tension of these ligaments and therefore the mediastinum is extremely important in enabling the diaphragm to move around these fixed points, more so than the muscle's pressure on the visceral mass.

MOTILITY

Lungs

One reason for our belief that visceral motility has a close relationship with embryogeny (page 9) is that the lungs, during the motility cycle, retrace their movement during fetal development. The lungs are the last important organs to appear; they can be perceived at the end of the second month and from then on grow very quickly. They are originally posterior but move anteriorly on either side of the heart as they grow larger. After birth, when they have been expanded by the intake of air, their anterior aspects move entirely to the front. Likewise, motility is a pendulum-like motion with the lungs moving from a relatively posterior to a more anterior position.

The general motility of the lung is identical to its mobility, with a vertical axis of motion for the upper lobe and an obliquely inferolateral axis for the lower lobe. On the right, the middle lobe moves synergistically with the upper lobe. The axes coincide with the same structures, the bronchial trees. While "listening" for motility, it is easy to confirm by palpation the fact that the left bronchus comes off the trachea less obliquely than the right. Inspir is an external rotation of the structures and expir is a return to the original position as the structures move toward the midline.

Mediastinum

The motility of the heart itself cannot be felt, as it is obscured by the much greater motion (mobility) of the heartbeat. We are able to palpate the motility of the mediastinum, which appears to be the same as that of the sternum. The heart, being fixed, does not move; the upper part of the mediastinum, with the sternum, swings anteroinferiorly during inspir and posterosuperiorly during expir.

Indications for Visceral Evaluation

A decision to treat a thoracic viscera should follow an appropriate diagnosis, which leads to a reconstruction of the pathology of the illness. The difficulty in deciding whether thoracic viscera are involved in a problem results from the fact that, apart from the heart, the thoracic viscera are not vulnerable to pain. A problem with the esophagus is rarely the reason for pain in the thorax, unless there is vomiting or reflux.

A careful and thorough history and physical examination should help you decide when it is necessary to look for a problem in the lungs or mediastinum. The most common indications are:

- all sequelae of pulmonary or pleural illnesses
- cervicobrachial neuralgias
- neck pain, upper back pain or intercostal neuralgia
- gastroesophageal problems
- hepatobiliary problems

Obviously, pulmonary pathology can result in a wide range of disorders. Poor motility and/or mobility of the lungs can cause tension and retraction of these other tissues, leading to dysfunction.

These are indications to *look* for visceral problems within the thorax. You may find them, you may not. When no significant restrictions exist there, it is necessary to look elsewhere for the cause of the problem. We are in no way saying that all gastritis is caused by a lack of elasticity in the pulmonary parenchyma!

Evaluation

HISTORY

The goal of the history is to contribute to the development of a diagnosis from which we will decide whether or not we are able to treat the disease. It is the most important aspect of the evaluation if you have no other reliable means of approaching the patient. For those of us with the skills to listen to the tissues, the history can be either the most helpful or the least helpful part of the examination: helpful if it elaborates the etiology of the problem; not helpful if it obfuscates or distracts us from the problem. The history should begin with the trauma of birth and be conducted so as to elicit as complete a list as possible of any physical, viral, bacterial or psychological attacks the patient has suffered. In regard to the lungs, certain diseases such as tuberculosis remain taboo subjects even today; careful questioning is required to elicit a good history. Knowledge of the history of vaccinations and different treatments received is essential.

PHYSICAL EXAMINATION

A patient's blood pressure is of particular interest for our purposes in that there can be a *different systolic blood pressure in each arm* — varying by as much as 30mm Hg. We consider the higher reading to be the accurate one, as a restriction is likely to decrease blood flow. Reduced blood pressure in one arm often signifies an ipsilateral pleuropulmonary injury.

The radial pulse is important in many ways. Sternoclavicular compression can be tested for by exerting pressure on the sternoclavicular joint from below while taking the ipsilateral radial pulse. A decrease or disappearance of the pulse signifies a retraction of the underlying soft tissues. To test for thoracic inlet conditions, check for variations in the radial pulse as the ipsilateral arm is passively moved in abduction, extension and external rotation. A positive test (diminution of the pulse) points to a possible contraction of the thoracic inlet. Where rotation and sidebending of the cervical spine to the opposite side is added, this is called the Sotto-Hall test *(Illustration 2-12).* These motions of the cervical spine may also be done to the ipsilateral side (Adson's test). Changes in blood pressure and the radial pulse can occur with any pathology of the ipsilateral side.

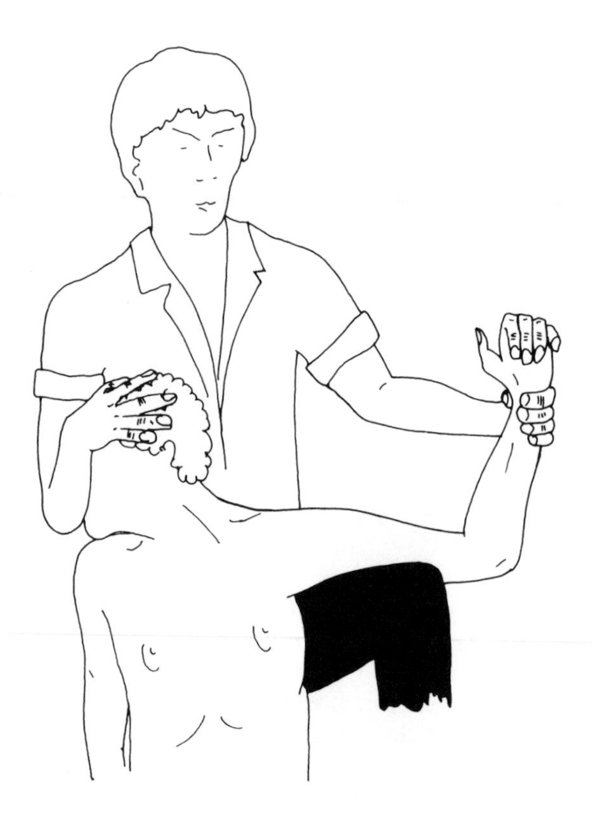

Illustration 2-12
The Sotto-Hall Test

The problem does not have to be immediately in the area of the thoracic inlet. For example, we have seen an impalpable pulse develop from a ptosis of the right kidney post partum. Similarly, during one experiment we observed a dramatic increase in the right radial pulse as measured by Doppler following manipulation of the liver.

Certain transient symptoms can be triggered by a general motion (e.g., sidebending with rotation to the opposite side) of the cervicothoracic region. If there is pain during the motion, the area of pain almost always represents the area with nerve root compression.

Auscultation and percussion are important parts of the physical examination. Look for all pleuropulmonary noises which reflect restrictions in ventilation; these generally result from a loss of elasticity of the pulmonary parenchyma. Auscultation and percussion will demonstrate the outline of the lungs and may also give information about problems with ventilation.

We do not believe that radiography is necessary in every case, but it should be used if there is any possibility of malignancy. Other examinations (e.g., lung perfusion scans) should be used when appropriate.

MOBILITY TESTS

The mobility tests serve to reveal any thoracic articular restrictions, i.e., of the thoracic vertebrae, ribs, sternum, or shoulder girdle. Mechanically, the rib cage and vertebral column are particularly important in regard to the functioning of the lungs. All articulations which involve these bones should be tested. We cannot detail all the mobility tests of this region. However, you should pay particular attention to those articulations which have close connections with the pleural dome and the cul-de-sacs; these tests are described below.

The first thoracic vertebra is tested in the seated patient by a movement which sidebends the cervicothoracic junction. You oscillate this area using a transverse motion with the greatest excursion at the head and T6, in such a way that T1 occupies the crossover point of the curves created *(Illustration 2-13)*. During the test, the index and third fingers are placed on either side of the spinous process of T1. Due to interapophyseal sliding, T1 conveys information on the opposite side from the movement: if T1 will not slide laterally toward one side, it is fixed on the opposite side.

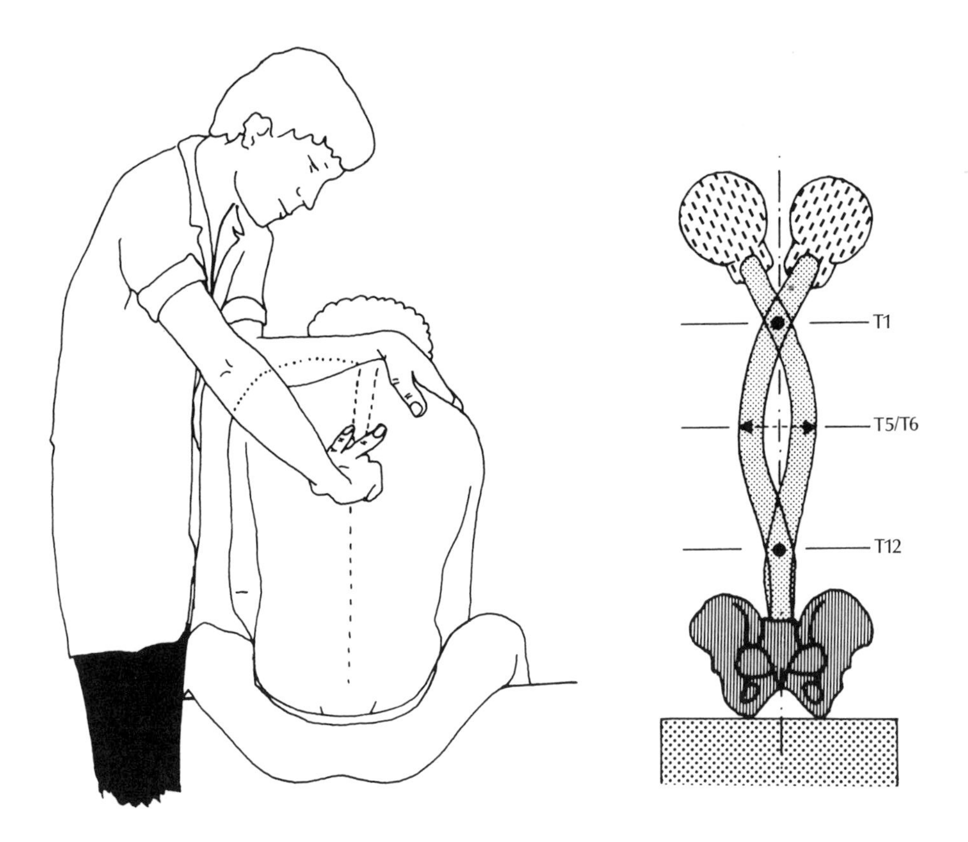

Illustration 2-13
Mobilization Test for Intervertebral Articulations

There are several methods of testing R1. It articulates with the manubrium and body of the sternum and transverse process of T1. We prefer to test costovertebral articulations with the patient seated. Place the radial edge of your index finger against the posterosuperior part of R1. Passively sidebend the cervicothoracic junction toward the side you are palpating, together with a rotation to the opposite side. The rib should easily move inferomedially; if it resists the pressure of the finger, it is restricted. The sternocostal articulation is tested in the supine position. Apply a light, springing movement directly to the area to evaluate the elasticity of the articulation. If it has a restriction, the tissues in this area will be less elastic.

The lower six costovertebral articulations are tested in the seated or lateral decubitus positions by rotating the thorax to the opposite side. Your fingers palpate the rib angles, which should move anterolaterally with the rotational movement. The lower costochondral articulations are tested by rotating the thorax to the ipsilateral side. The fingers palpate the anteromedial aspect of the ribs, which should move posterolaterally with the rotational movement.

The test for T11/12 is similar to that for T1. In order to focus the motion at the level of T11/12, the apices should correspond with T5/6 and the pelvis *(Illustration 2-13).* An easy interapophyseal sliding movement is a sign of a good physiology of intervertebral articulations.

MOTILITY TESTS

These tests are done in the supine position. As the motions being tested for are quite subtle, it is important that both you and the patient be quiet and relaxed. In all these tests, it is a good idea to compare the motions of inspir and expir with those of

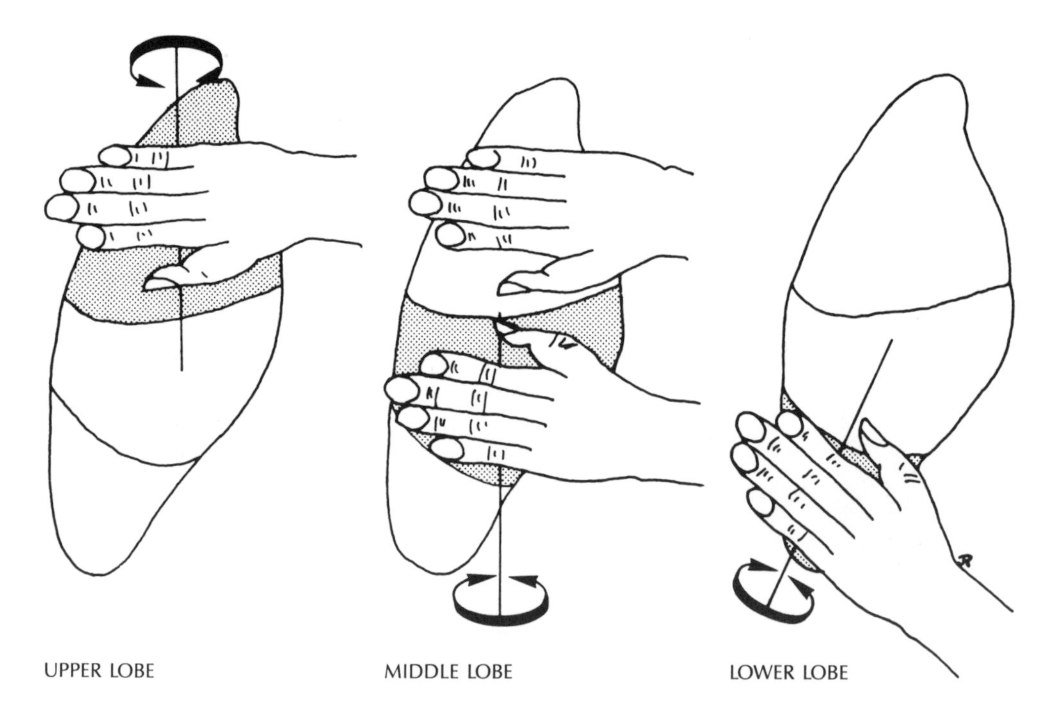

Illustration 2-14
Motility Tests for the Right Lung

inhalation and exhalation. The rhythm of motility should be 7-8 cycles per minute, which is roughly half that of the respiratory rhythm. The frequencies of respiratory rhythm and motility are different and not in sync with each other. Be relaxed when you "listen" to motility; it is relatively easy to ignore respiration as you do this.

In the right lung, the test for the upper lobe consists of feeling the horizontal rotation of the lobe around its apical bronchus. The middle lobe has the same motility as the upper lobe. To isolate it, it helps to inhibit the upper lobe by pressing on it gently. The oblique nature of the lower lobe's axis of rotation should be appreciated *(Illustration 2-14)*.

The tests for the left lung are identical to those for the right lung, except that there is no middle lobe. The upper lobe seems to rotate around the segmentary apical bronchus. The lower lobe also rotates around the bronchial tree. On the left, the axis makes a larger angle with the vertical than on the right.

The aim of the test for the mediastinum is to feel for the anteroinferior swinging motion of the superior mediastinal area and sternum during inspir, and the opposite during expir *(Illustration 2-15)*.

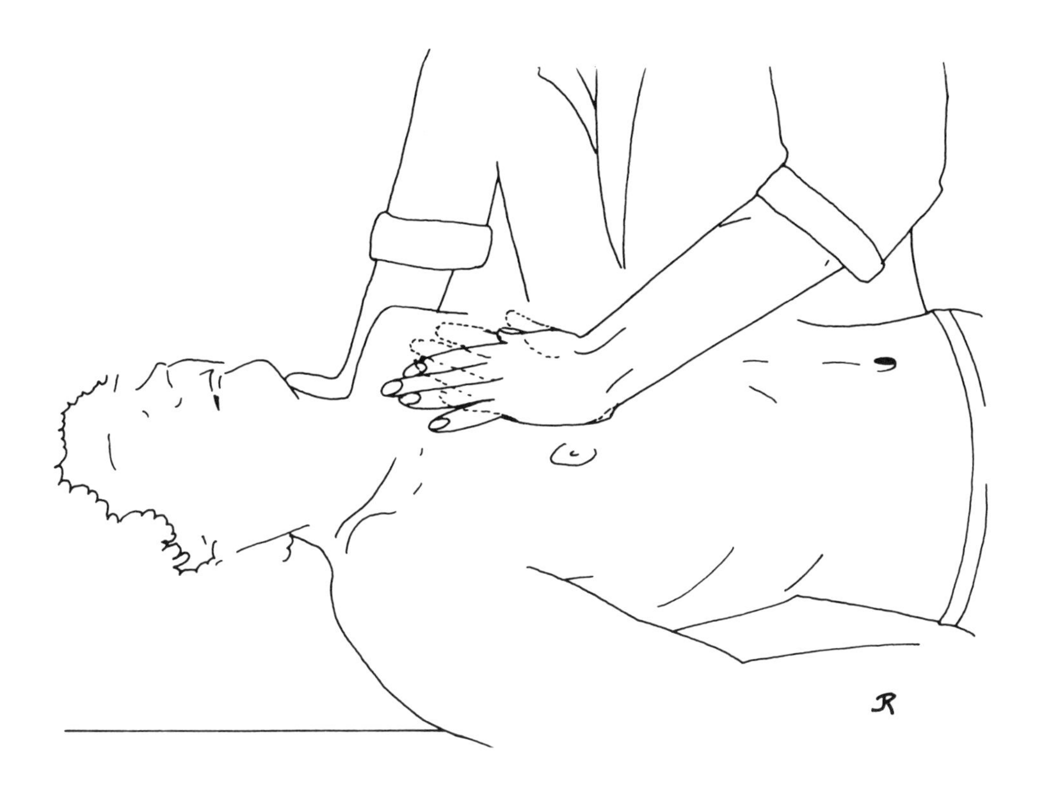

Illustration 2-15
Motility Test for the Sternum

Restrictions

VISCERAL ARTICULAR RESTRICTIONS

These are pleural adhesions and are quite common. In 1852, Cruveilhier wrote: "Let us say that it is very rare to find lungs without any adhesions on the surface. The ancients considered fibrous or other adhesions as natural adhesions." We do not believe that mankind today has pleura of a superior quality.

These adhesions are found in the areas of least mobility. The cul-de-sacs tend to become obliterated during deep inhalation. Because of the fissures, the lobes slide one upon the other, particularly during forced inhalation. The patient with a sedentary occupation, who does not exercise regularly, has a good chance of having adhesions in these places. The pleural dome is an area of minimal interpleural sliding. Adhesions of the pleural dome, which are always associated with a ligamentous restriction of the suspensory ligament, will be discussed below. In general, adhesions become the new axes of motility.

The most lateral part of the costodiaphragmatic cul-de-sac is the deepest. The pleura have almost no possibility of sliding upon each other here. An adhesion in this area becomes the new center of motility. When there is a problem here, the motility of the lower lobe involves rotation around an anteroposterior axis *(Illustration 2-16).*

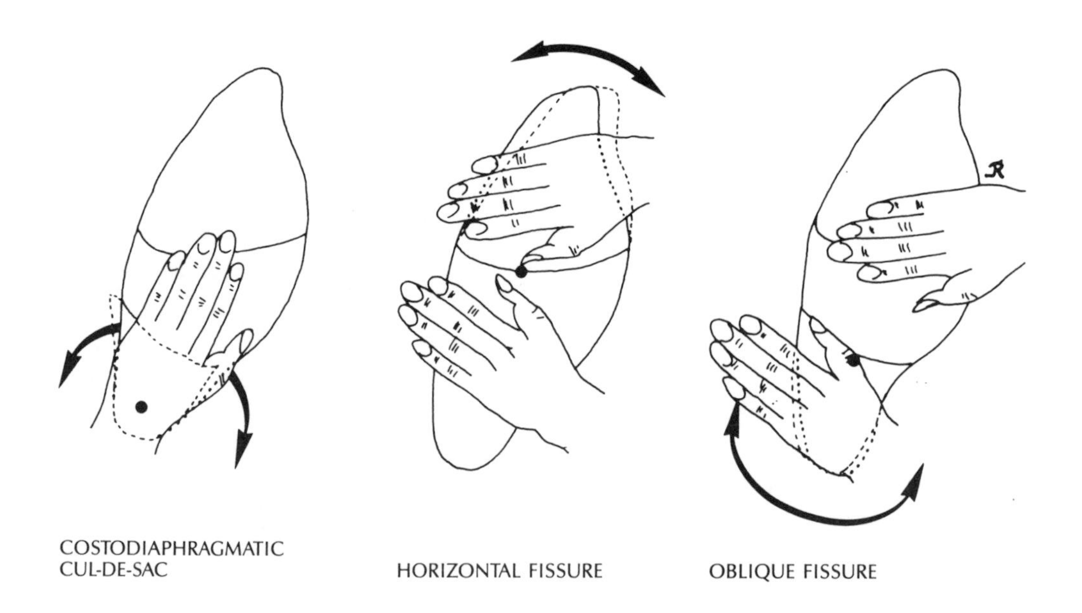

Illustration 2-16
Restrictions in Motility of the Right Lung

On the right, because of the horizontal fissure, it is easy to inhibit the motion of the middle and lower lobes with the left hand and test the upper lobe with the right hand.

Depending on the position of the patient and whether you are right or lefthanded, the procedure may be reversed (i.e., inhibit the upper lobe and test the middle lobe). With an adhesion of the horizontal fissure, the motility of the upper lobe is changed into one of frontal rotation around an anteroposterior axis which passes through the adhesion *(Illustration 2-16).*

The oblique fissure is treated in the same manner as the horizontal fissure. It is necessary to inhibit the motion of the upper and middle lobes on the right, or the upper lobe on the left. With an adhesion of the oblique fissure, the motility of the lower lobe is changed to one of frontal rotation around the adhesion *(Illustration 2-16).*

VISCERAL LIGAMENTOUS RESTRICTIONS

The ligamentous system of the thoracic cavity consists of the interpleural ligaments and the various ligaments which stabilize the heart (page 40). This system is reinforced laterally and inferiorly by the pulmonary hili and lung ligaments. It renders the constituents of the mediastinum interdependent *(Illustration 2-17).*

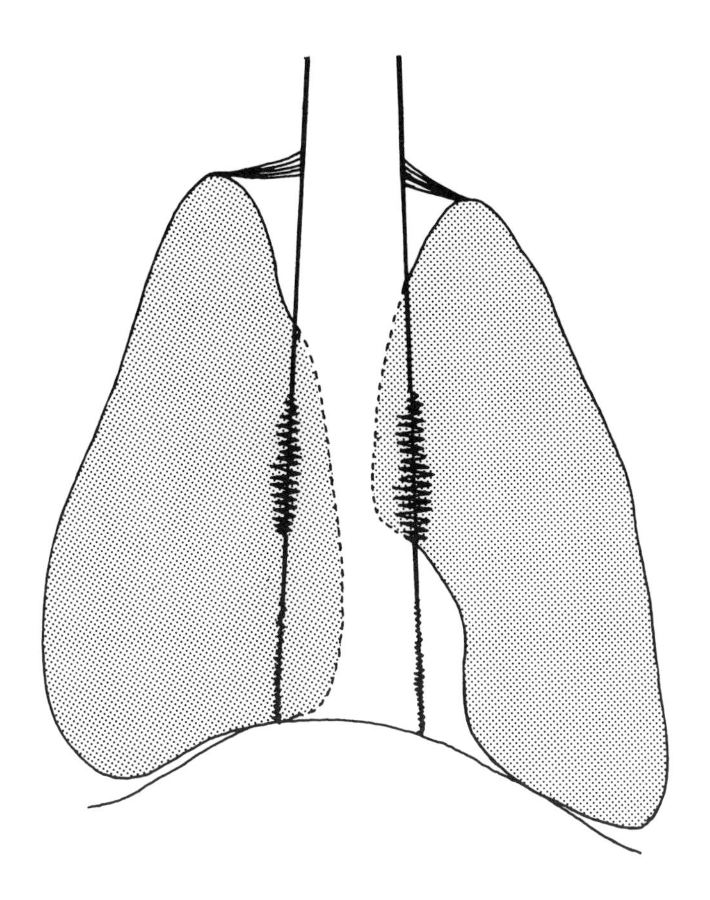

Illustration 2-17
Schematic Representation of Mediastinal Restrictions

The mediastinum functions as a midsagittal diaphragm or interpleural ligament. During respiration, the mediastinal viscera exert traction on the ligamentous system. This traction is exerted bilaterally during inhalation. Physiologically, the traction exerted on the right is identical to that on the left and they should cancel each other out. This mechanism maintains the mediastinum in place on a frontal plane.

RESTRICTIONS OF THE MEDIASTINUM

When the pulmonary parenchyma loses its elasticity, the traction of the muscles of inhalation is not cushioned by pulmonary expansion, but passes directly to the wall of the mediastinum. The traction is stronger on the injured side and the mediastinum will therefore be deviated toward this side. The anterior and superior regions of least resistance of the mediastinum are the most likely to be affected in this way.

This deviation can be a subtle manifestation, only noticeable when listening, or can be serious enough to have a structural effect wherein the affected atrophied lung lets the other lung cross the median line.

The ligamentous system, from which the phrenic center is suspended, is drawn downward with each inhalation. The sinewy fibers of the phrenic center can become fibrosed (a chronic disease), or be submitted to acute pressure by a hypertonic diaphragm. This latter condition is the only thoracic visceral restriction which causes painful symptoms or localized discomfort. With an acute pressure on the mediastinum, we find symptoms of recurrent spasms with sensations of pulling at the throat, cardiac knots, suffocation, etc.

RESTRICTIONS OF THE SUSPENSORY LIGAMENT

This ligament is made up primarily of sinewy fibers, but there are also some muscular fibers derived from the scalenes. These fibers form a net which encloses each

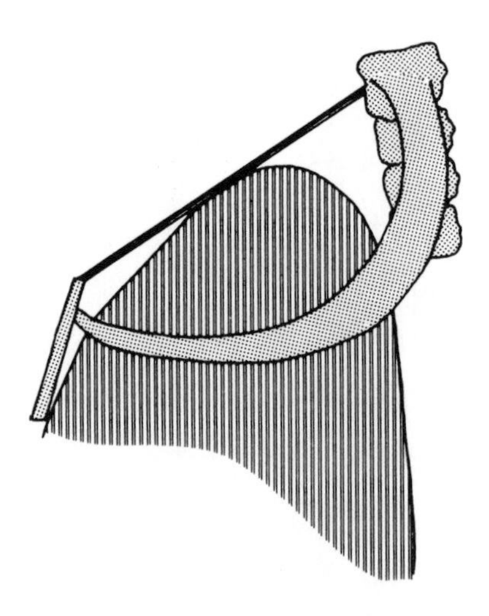

Illustration 2-18
Schematic Representation of the Suspensory Ligament of the Pleural Dome

hemithorax at the upper part and inserts tangentially into the intrathoracic fascia of the pleural dome *(Illustration 2-18)*. This upper thoracic region is connected with the scapular girdle and the cervicothoracic vertebral junction, and is very mobile.

There are many causes of rigidity in this area, particularly localized mechanical causes; any untreated restrictions of T1, R1, or the clavicle would bring about fibrosis in the upper diaphragm. The motility of the upper lobe is altered by this type of restriction. With listening, the motion would be felt as a frontal rotation around the apex, indicating a pleural adhesion at the top of the lung *(Illustration 2-19)*. Restrictions of the pleural dome can also be caused by a loss of elasticity at the top of the lung. Sclerosis of the parenchyma causes a kind of retraction which can lead to a deviation of the upper part of the mediastinum toward the injured side.

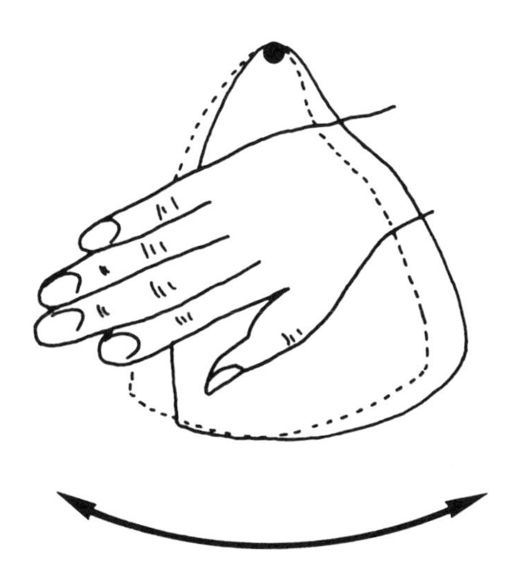

Illustration 2-19
Pleural Dome Restriction

If the function of an organ or part of an organ is lost, the self-healing reaction of the organism will often replace it. For example, when an upper lobe loses its elasticity, the space which is thereby freed will be filled up by the other lobes.

Faced with a complex of pleural adhesions and ligamentous restrictions, it may be difficult to decide which is the most important. Experience has shown us that one should place primary importance on the ligamentous system; treatment here will lead to improvements at every level.

Manipulations

As an osteopath, you know that it is not possible to give a universal standard treatment for a particular type of problem. The treatment will be adapted to the particular diagnosis and patient. We shall therefore not describe specific treatments, but therapeutic techniques which can be combined to give you a variety of possible treatments. Disorders caused primarily by restrictions in the motion of the lungs, pleura or mediastinum can be treated by direct, induction or combined techniques (see chapter 1).

DIRECT TECHNIQUES

These are used for problems of mobility and are generally performed with the patient seated. For the lungs, direct techniques often fall under the category of "general stretching" because they involve large parts of the body, but the stretching can be focused on a small area, in a very precise manner.

The fibers of the suspensory ligament run anteroinferiorly. Stretching the ligament requires moving the head and neck away from the affected side of the thorax. For example, for a problem of the right pleural dome the patient's head is rotated to the left to stretch the scalene muscles. The stretching effect is accentuated by adding left sidebending of the head and also by pushing inferomedially on the upper part of the thorax on the affected side *(Illustration 2-20)*. The stretching is applied gently and rhythmically, until a release is felt.

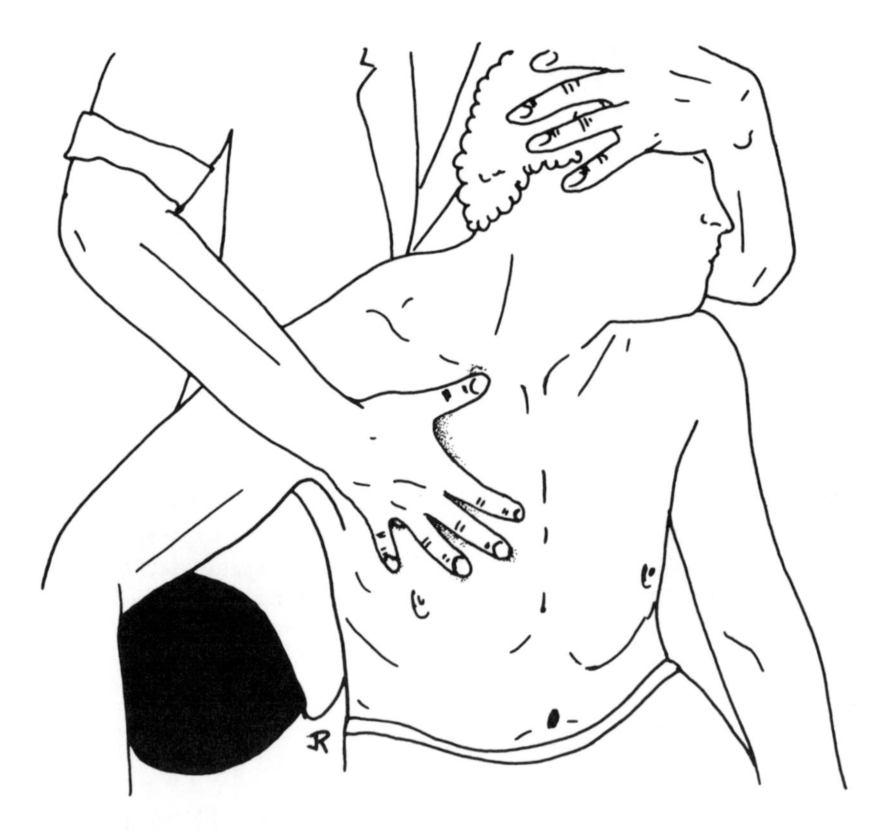

Illustration 2-20
Stretching of the Suspensory Ligament of the Pleural Dome

By varying the relative degree of flexion/extension, rotation and sidebending you exert on the thorax, it is possible to localize your stretching to the desired part of the parietal pleura. For example, to stretch the parietal pleura in the area of the right nipple, have the patient put his right hand behind his head. Then press the right craniocervical junction superiorly and rotate it to the left (to release the posterior attachments of the pleura) or right (for the anterior attachments). At the same time, press on the lateral aspect of the left ribs to sidebend the thorax to the left *(Illustration 2-21).*

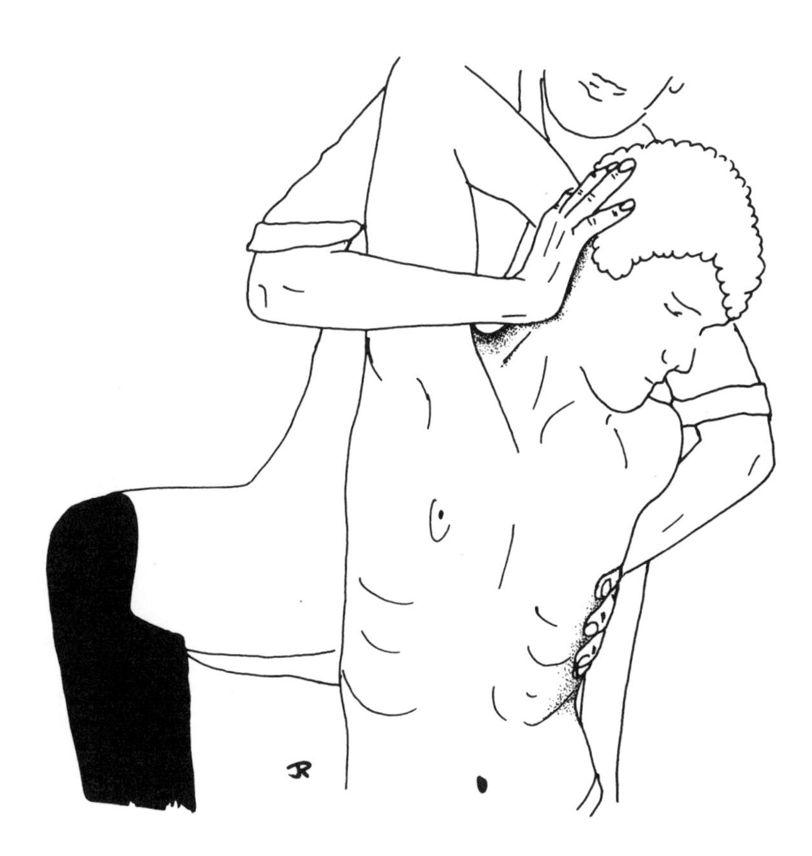

Illustration 2-21
Stretching of the Parietal Pleura

INDUCTION TECHNIQUES

With induction, it is essential to encourage or exaggerate the motion, or part of the motility cycle, that is more prominent. We call this going in the direction of the lesion, or going "into" the lesion. It is not our action per se which is therapeutic, but rather the patient's response. We do *not* attack the organ; we stimulate it so as to focus and accentuate its own inherent healing forces.

Simple induction is performed by following the motility of the underlying tissues. When the motility is decreased or its axis of motion changed, you first follow passively (listening) and then progressively induce or encourage it to follow the direction in which it goes most easily. If there is a change in the rhythm, follow it. After several successive cycles of inspir and expir, you often will be able to feel a "still point" at the endpoint of its motion of ease. If this motion does not return with a more physiological axis and amplitude within approximately half a minute, restart it by *gently* encouraging the motion in the direction it went most easily before the onset of the still point. Follow the motion of motility passively and, if it is not completely normalized, repeat the induction procedure just described.

In general, the sequence is:

Listening ► induction ► still point ► induction ► listening

Sometimes, motility will improve without occurrence of a still point. When this happens, simply continue the process until on listening the motility appears normal. The still point is not essential.

Induction with counter-traction differs from simple induction in that you inhibit the area(s) contiguous to the area being treated. Apply a gentle pressure (slightly more forceful than that applied for listening and induction) to temporarily decrease its motion, which will enable you to more easily focus your attention on the area. In the thorax, this technique is most useful for problems with a fissure or the mediastinum.

For example, to treat a restriction of the horizontal fissure, inhibit the middle and lower lobes with one hand. With the other hand, listen to the upper lobe, induce the motion just to the still point, feel the release and then induce again, as necessary, until normal motility is restored *(Illustration 2-22).*

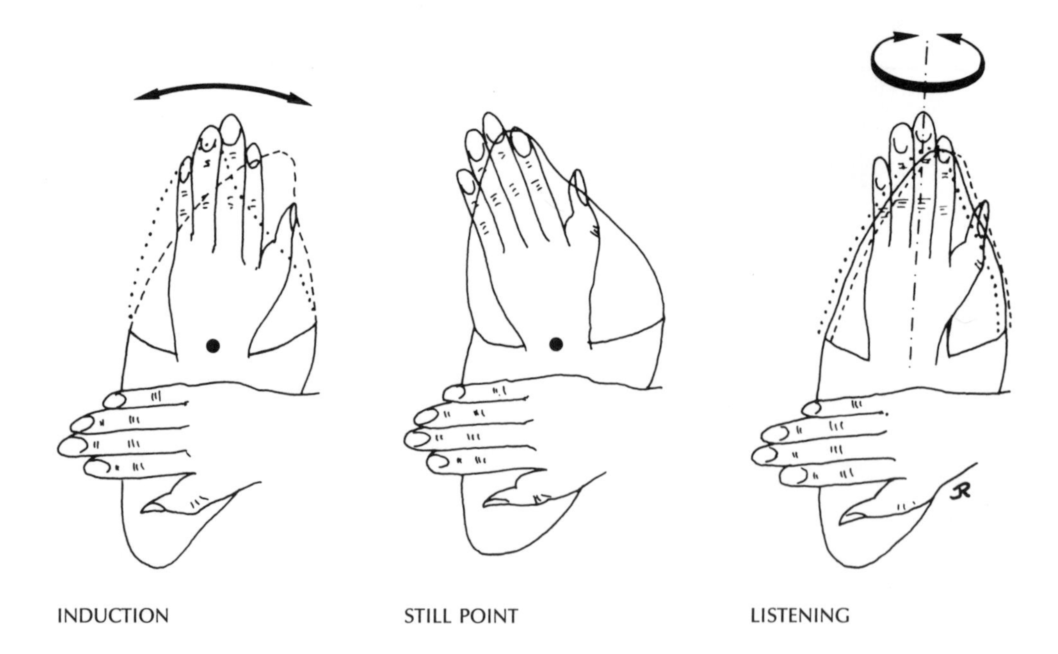

Illustration 2-22
Induction of the Superior Lobe with Counter-Traction

For a restriction of the oblique fissure, inhibit the upper and middle lobes and induce the lower lobe. Then:

- Listen to the dysfunctional motility. If necessary, exert a slight upward traction in the direction of the apical segmental bronchus. After several cycles the still point is felt, usually at the end of inspir
- Feel for the release and then induce the motion of normal motility. After several inspir/expir cycles, go back to listening; the superior lobe should now rotate around a vertical axis
- If the motion is still disturbed, repeat the procedure.

This technique is essential in treatment of adhesions of the fissures. The counter-traction allows a slight separation to take place in the fissure between the two hands and also permits the isolation and accentuation of the pathological pendular motion, the center of which is the adhesion.

COMBINED TECHNIQUES

In combined techniques for the thorax, the long lever arm is used to reduce local pressure; the short lever arm is used for induction. This technique is used primarily for

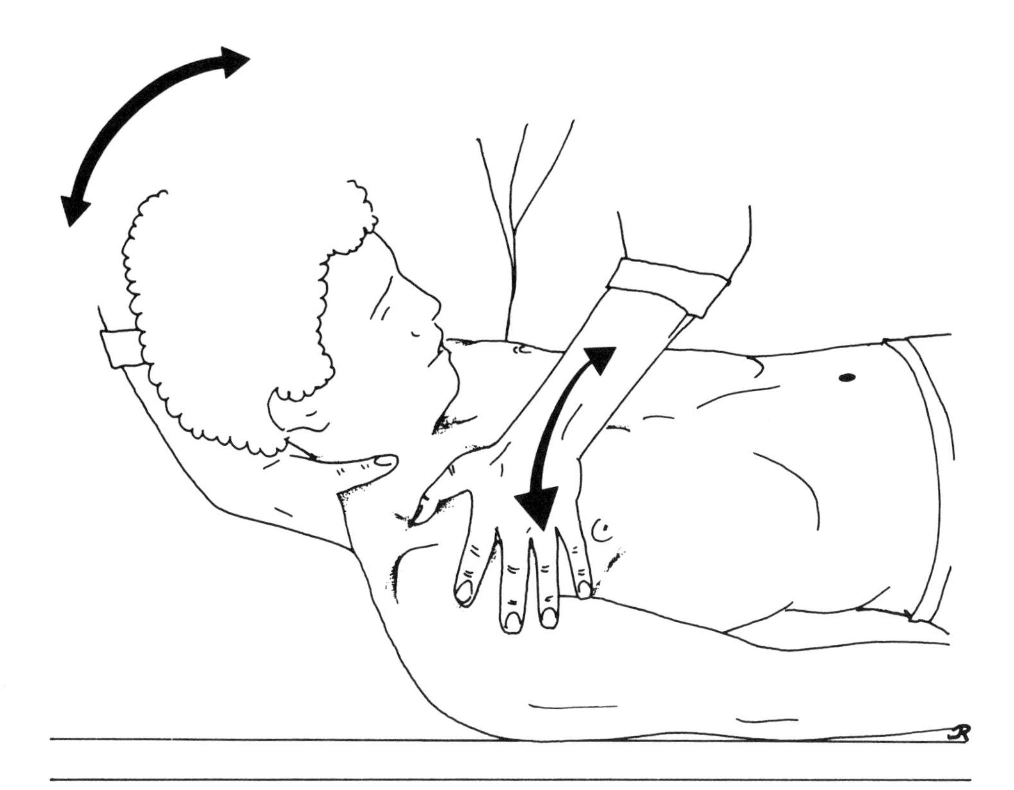

Illustration 2-23
Combined Manipulation of the Pleural Dome

restrictions of the suspensory ligament *(Illustration 2-23)* and the mediastinum *(Illustration 2-24)*, which are often functionally related. The long lever arm can be used to release local tension and, in both cases, the vertebral column should be bent forward to the level of T5 to focus the effect.

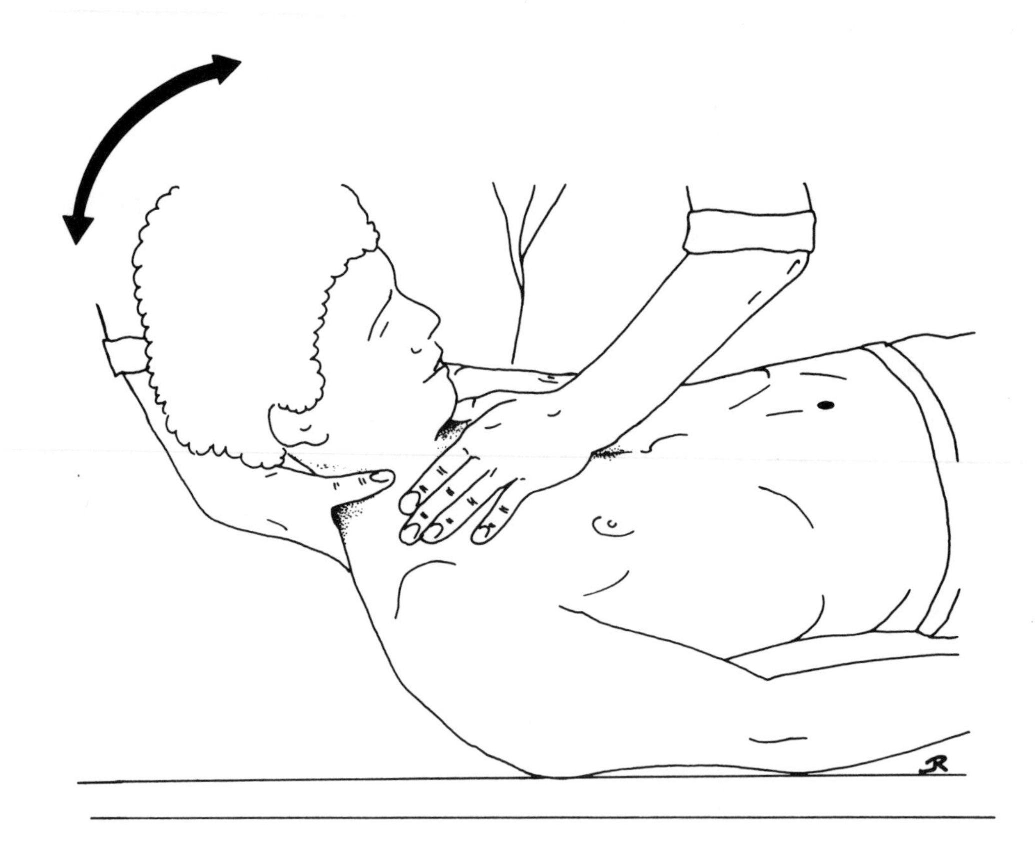

Illustration 2-24
Combined Manipulation of the Mediastinum

The short lever arm induces the motion. In most cases, the lesion will be exaggerated during the inspir motion. The long lever arm, in sync with motility, brings the vertebral column into flexion during inspir and returns it to a normal position during expir. If the lesion is exaggerated during expir, we usually flex the vertebral column during that phase and return it to a normal position during inspir.

The still point is usually found with the column bent in an inspir position. Hold this position until a release is felt. Then induce normal motion for a few cycles. At this time, listen to judge the quality of the motility. If it is not satisfactory, repeat the technique.

When the fasciae, pericardial ligaments or lung ligaments are fibrosed or "shortened," the phrenic center, which is suspended from them, no longer has the necessary latitude for a qualitatively normal respiratory motion. The vertical tension is further aggravated by inhalation, the diaphragm is permanently tensed and a mediastinophrenic node exists, which should be broken up.

After performing the combined technique described above, you can free the phrenic center using the following technique. This employs both long and short lever arms, but does not involve induction. The patient sits facing away from you, allowing you to regulate the degree of vertebral flexion. Usually, the vertebral column is totally bent to prevent the mediastinum and abdomen from becoming tense. Slide the ulnar edges of your hands under the thorax, at the level of the costal cartilages, against the diaphragm. Maintain this gentle, but firm, bimanual pressure and then mobilize the thorax using the arms and back in a rhythmic and gentle, but firm, manner. When the area under the hands releases, move your hands laterally on both sides to cover all the costal insertions of the

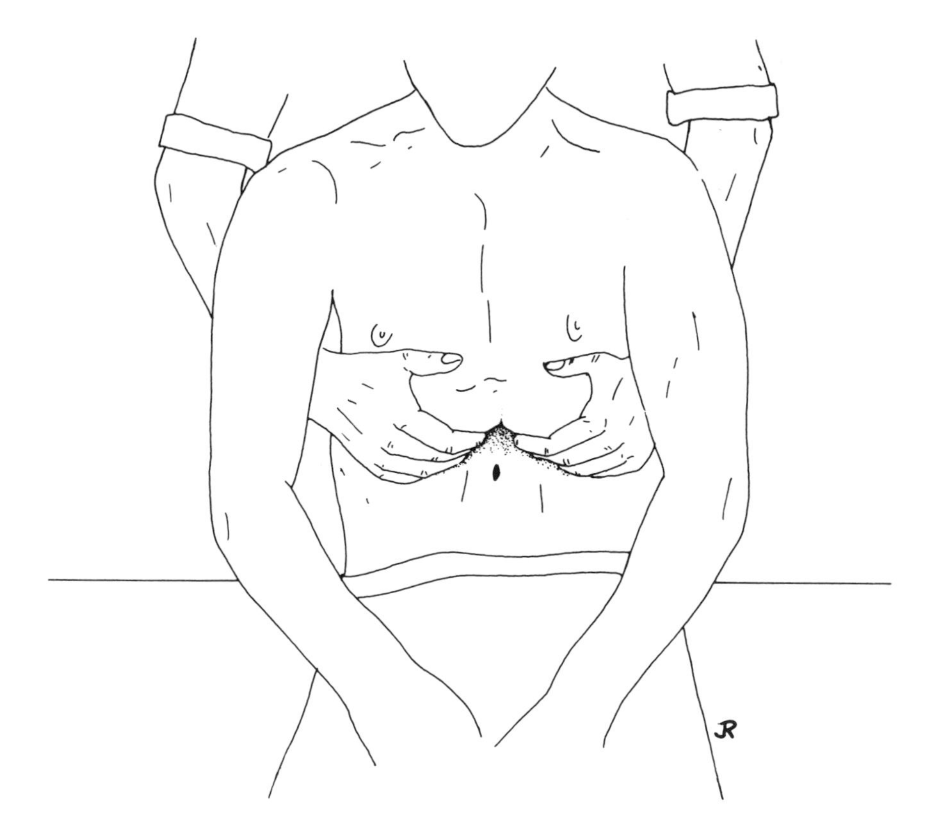

Illustration 2-25
Combined Manipulation of the Phrenic Center — Seated Position

diaphragm *(Illustration 2-25)*. Sometimes, you may immobilize the patient in the position of maximum balanced tension and ask her to breathe in. In this way, the diaphragm corrects itself.

This technique can also be done in the supine position, with the patient's hips and knees flexed and a large cushion placed under the nape of the neck. One of your hands is in contact with the diaphragm, while the other hand adjusts the position of the knees for maximal effect *(Illustration 2-26).*

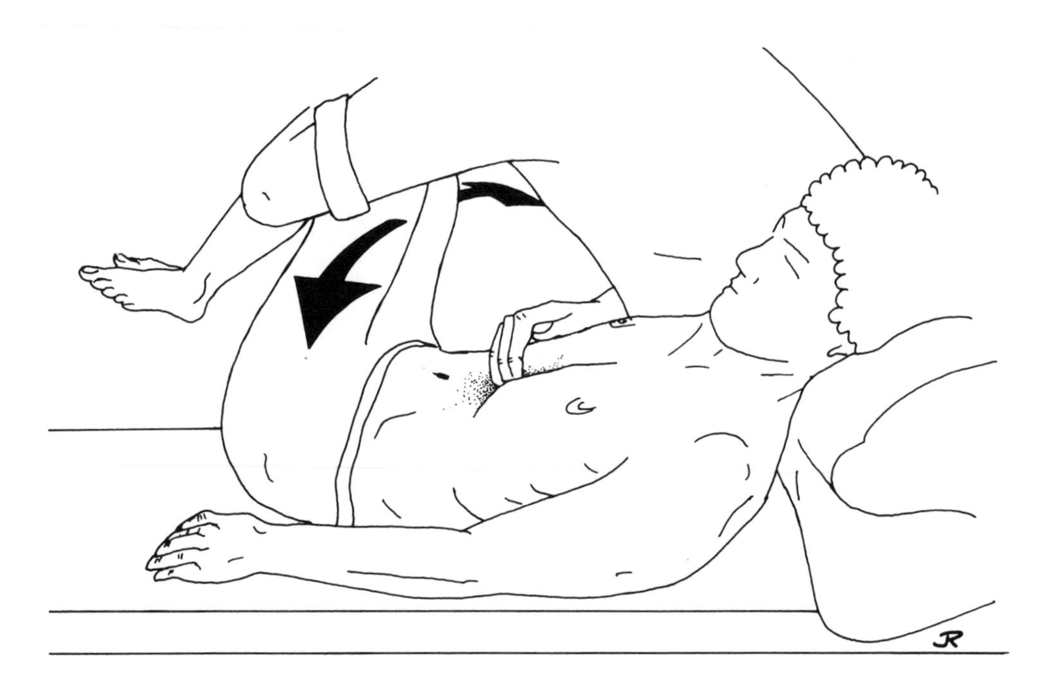

Illustration 2-26
Combined Manipulation of the Phrenic Center — Supine Position

Effects

The effect which is sought and accomplished by these intrathoracic visceral techniques is quite simple: harmony. "Harmony" is a key word in osteopathy. The techniques also restore a proper relationship between the contents and the containers of this area. The end result is a return of vitality, mobility and dynamism to the viscera.

These techniques will not restore the damaged part to its pristine state. Retraction of the injured pulmonary parenchyma will persist, as will the scarring pleural adhesions. However, by increasing the motility of the injured organ, you improve its function, eliminate secondary problems and help compensate for the local deficit (pages 28-29). All lung and pleural illnesses can and should be treated by these techniques, with the exception of the acute stages of serious pneumonias which require antibiotics or other measures. We have used mediastinal induction (with or without counter-traction) on children, with rapid results. This is based on our finding that both upper and lower respiratory tract infections drastically decrease the motility of the mediastinum. It is also possible that this technique stimulates the thymus in children.

Adjunctive Considerations

ASSOCIATED RESTRICTIONS

Deviations of the mediastinum can cause a chain reaction of functional problems. Whether this be a lateral deviation or a vertical retraction, diaphragmatic movements will be disturbed. The phrenic center is drawn upward. The liver and stomach follow the movement of the diaphragm and the height of the lungs is reduced. Because the diaphragm and mediastinum are already under pressure, respiratory excursion will bereduced. With lateral deviations, the esophagus will also be affected.

Deviations or retractions of the mediastinum can also cause problems in the abdominal region. In chapter 1, we discussed how the abdominal viscera, in connection with the diaphragmatic muscle, are always drawn upward by the relatively lower pressure of the thoracic cavity. Changes in the forces focused on the mediastinum cause changes in those focused on the diaphragm, which are then transmitted to the liver and stomach.

If thoracic mobility and motility are abnormal, the first signs of injury usually manifest as osteoarticular restrictions. These may involve articulations of the ribs, vertebrae, sternum, costal cartilage, etc. We have seen how even small abnormal motions, repeated millions of times, can bring about skeletal deformations of the thorax.

Restriction of the suspensory ligament can cause destabilization of the cervicothoracic junction. This vertebral restriction can cause problems such as cervicobrachial syndromes or intercostal neuralgia. Restriction of the suspensory ligament can also affect the autonomic nervous system because of the connection between the head of R1 and the inferior cervical (stellate) ganglion.

Intervertebral and costovertebral restrictions can be consequences of the deterioration of the reflex arcs linked to injured viscera. The abnormal afferent nerve impulses coming from the viscera produce a modification of the efferent nerve impulses in the same dermatome and elsewhere. Often this reaction is not enough to solve the problem. The soft tissues connected with the vertebral articulation, which are thereby bombarded by aberrant nerve impulses, react by becoming hypertonic, leading to restriction of the articulation. These reflex restrictions are often found at the level of the T1-4. It is not possible to state that one segment is more important than another. Quite often the restriction begins at the level of the costovertebral articulation.

Restrictions of the suspensory ligament sometimes causes contraction of the thoracic inlet and compression of the subclavian artery. This is manifested by a disturbance of the radial pulse on that side.

The connections between osteoarticular restrictions and visceral lesions are either purely mechanical or reflexive in nature. Each type of problem should be treated differently. If the connection is mechanical, treat the osteoarticular restriction first. Once this is released, treatment of the viscera by induction and stretching can begin and will give excellent results.

In general, do not bother to treat reflexive articular restrictions directly, as they will disappear automatically after proper treatment of the viscera. In chronic cases, the reflexive vertebral restriction can be so severe that it becomes a kind of "primary mechanical restriction." When this happens, it is necessary to treat the osteoarticular restriction first, as otherwise it will be a factor in the continuation of the visceral restriction.

In conclusion, we would like to emphasize two essential points:

- Pathologies of the pulmonary parenchyma, or pleura, can produce osteoarticular restrictions of the thorax and cervical vertebrae.
- Such restrictions do not necessarily signify an injury of the thoracic viscera.

Do not fall into a mechanistic, reductionist trap of seeing all medical problems as due to one or two causes. We are incredibly complex organisms and the health practitioner must always attempt to see patients clearly as they truly are — not fit them to a Procrustean bed based on the practitioner's limited understanding.

RECOMMENDATIONS

Everyone has minor pleural adhesions in the same way that they have minor vertebral restrictions. A scarring adhesion may follow an old pathology, known or unknown, benign or severe. Depending upon the vitality of the subject, it can be kept latent and do no damage. Because the lung feels no pain, the patient will consult you for secondary problems which bring about uncomfortable symptoms. During your diagnosis, you should systematically examine the thorax and its contents.

We urge you to try to attract the children of your patients for consultations. A disturbance in motility is the first clinical sign of a visceral lesion. The first diseases children catch are carried in the air and affect the respiratory organs. By treating the young patient at an early age, you help him avoid the sequelae of pleuropulmonary illnesses, which can cause so much trouble in adult life.

Chapter Three: The Abdominopelvic Cavity

Table of Contents

CHAPTER THREE

The Abdominopelvic Cavity

As a practitioner of manipulative medicine, you are aware of the marvels and complexity of the human body, including the mechanics of the musculoskeletal system, the physiology of each of its systems and their correlations, and the power and intricacy of the immunological reactions. All of these are important in understanding how to perform visceral manipulation. Before focusing on individual abdominal viscera, we would like to review the general anatomy and mechanics of the abdomen and its contents.

Anatomy

Schematically, the abdomen can be viewed as a vertical semicylindrical container with an upper base (the diaphragm); a lower base (the pelvis) which has a hole in it closed by the perineum; posterior and lateral walls made up of bones (lumbar column, lower ribs and iliac crest) and short, thick muscles; and a comparatively thin anterior wall made up of muscles.

The abdomen contains three types of organs: intraperitoneal, retroperitoneal and pelvic. Most of these organs are compressed by the diaphragmatic piston 15-18 times per minute. Because the abdominal contents always occupy the same volume, it is the anterior muscular wall which is passively distended under the pressure of the diaphragm.

INTRAPERITONEAL ORGANS

This group of organs (stomach, liver, spleen, jejunum, ileum, transverse and sigmoid colons) is enclosed in the parietal peritoneum, which cannot be enlarged but can be deformed. Most of these organs are of a semifluid consistency. Each intervisceral space is a virtual cavity of minimal volume containing peritoneal fluid, which prevents the viscera from adhering to each other. Pressure between viscera is much lower than that inside a viscera. The mechanism of intracavitary pressure (pages 15-16) causes the viscera to move together, sliding one against another, without ever sticking to each other.

RETROPERITONEAL ORGANS

These organs (kidneys, adrenal glands, pancreas, duodenum, ascending and descending colons) are located behind the parietal peritoneum, which is a supple structure, and in front of the posterior musculoskeletal wall. The column of intraperitoneal viscera is held in place by the tonicity of the abdominal muscles. Because of the position of the column, it has a tendency to fall forward, and thus the anterior abdominal muscles need to be the most active. Since this column is homogenous and compact, almost all the pressure exerted by the anterior abdominal muscles on these viscera is transferred to the posterior parietal peritoneum, perirenal fascia and kidneys. The muscular pressure is transferred posteriorly, and the kidneys are held in place, mainly because of intracavitary pressure and turgor.

PELVIC ORGANS

These organs (bladder, rectum, uterus, etc.) are found underneath the column of intraperitoneal viscera and fill up the pelvic basin. At first glance, they appear to be under great pressure from the viscera above. Actually, the pelvic inlet inclines more or less forward and the viscera themselves have a dome form which is convex superiorly, giving them a structural power similar to that of a cathedral. Most of the pressure of the abdominal column is exerted upon the internal iliac fossae and the ischiopubic rami. Because of their domelike form, the pelvic viscera are able to distribute the remaining pressure onto the perineum without being squashed. The perineum functions to some extent as a shock absorber of this residual pressure, which is minimal compared to the original pressure.

Abdominal turgor is essential for proper function of the pelvic organs. Because of turgor, the abdominal column functions as a solid mass resting upon skeletal elements of the pelvis. If this mass were soft and liquid, the pelvic viscera would be compressed. Adequate tone of the anterior abdominal musculature is an essential part of the turgor effect.

PERITONEUM

Of all serous membranes, the peritoneum is the largest and most complex. Its parietal surface has an area of approximately $2m^2$. There are approximately 50ml of serous fluid between the parietal and visceral layers, acting as a lubricant for the intraperitoneal organs. Inflammation can result in increased secretion of this fluid.

The parietal peritoneum lines the abdominal wall, has sensory innervation and is more solid than the visceral layer. It is thicker in the iliolumbar area and lined on its deep side with cellular subperitoneal tissue.

The visceral peritoneum, which loosely covers the viscera and lacks innervation, is deep to the parietal layer. It is a thin, transparent, very elastic layer, through which one can see the color of the organs. Except in the case of the liver and spleen, it is not attached to the organs.

Peritoneal Cavity

This is the space between the two layers. It is a virtual cavity and the pressure inside it is so much lower than that inside the nearby organs that the two layers are always "hunting" the largest area of contact between them. Because of the peritoneal fluid, and because the organs are in perpetual motion due to the action of the diaphragm, no adhesions normally develop between the two layers. The peritoneal cavity is enclosed all around, except in women, where it communicates with the uterine tubes via the abdominal ostia (see chapter 10). It must be emphasized that communication across these two serous membranes is very rare. The lowest part of the peritoneal cavity, located between the rectum and posterior uterus, is called Douglas' pouch (or rectouterine pouch).

The peritoneal cavity is divided into two portions by the mesocolon. The supramesocolic portion encloses the liver, stomach, pancreas and spleen. It is bounded anteriorly by the abdominal wall, posteriorly by the thoracolumbar wall, superiorly by the diaphragm and inferiorly by the transverse mesocolon and the two phrenicocolic ligaments. At the anterior edge of the transverse mesocolon, it communicates with the rest of the abdominal cavity. The gastrohepatic omentum divides this area into three secondary cavities: the hepatic fossa, gastric fossa and posterior cavity of the omenta. The hepatic fossa communicates with the parietocolic area on the right and the gastric fossa with that on the left.

In contrast to the other two cavities, the posterior cavity of the omenta is quite isolated from the peritoneal cavity. It communicates with the hepatic and gastric fossae via the epiploic foramen, which is an oval orifice limited posteriorly by the inferior vena cava, anteriorly by the hilum of the liver, superiorly by the caudate lobe and inferiorly by the superior duodenum. The posterior cavity is a space in which the stomach can slide. It is bounded anteriorly by the lesser omentum and stomach, inferiorly by the greater omentum and transverse colon and posteriorly by the transverse mesocolon, pancreas, liver and (on the left) spleen.

The submesocolic portion of the peritoneal cavity is bounded superiorly by the transverse mesocolon and transverse colon, inferiorly by the pelvis and elsewhere by the abdominal wall. This portion can be further subdivided into two mesenterocolic spaces: right (between the right side of the mesentery and the ascending colon) and left (between the left side of the mesentery and the descending colon).

Vascularization and Innervation

The peritoneum does not have its own vascular system; it relies on vessels of the nearby organs. It does have its own lymphatic vessels. Most of its innervation comes from the lumbar and solar plexuses. Serious reflex phenomena can be caused by an irritation of the parietal peritoneum — these reflexes can affect cardiac, respiratory, renal and intestinal functions and are therefore dreaded by many surgeons.

Physiologic Motion

ABDOMINOTHORACIC RELATIONSHIP

The mobility of the intraperitoneal viscera is under the control of physical laws which apply to the pressure of fluids and gases. The pressure of the abdominal cavity is noticeably higher than that of the thoracic cavity. It seems that the thoracic cavity attracts the abdominal cavity, and similarly that the abdominal viscera are permanently drawn up by the diaphragm (page 7). This thoracic "drawing up" takes place because the diaphragm, which is a supple structure, offers a contiguous elastic connection between the two cavities. Its dome shape is a witness to the suction effect of the thoracic cavity. The peritoneum, which is attached to the diaphragm, can only follow.

RELATIONSHIPS BETWEEN THE ABDOMINAL VISCERA

The pressure in the abdominal cavity is lower than that of the viscera. The viscera contact each other as much as possible, taking up only a small volume considering their number. It is this phenomenon which makes the peritoneal cavity only a virtual one. Although the abdominal viscera enclosed in the peritoneum are of different shapes and composition, they are encircled by a muscular belt and form a relatively homogenous column of viscera, due to intracavitary pressure and turgor. Thus, the thoracic "drawing up" is transmitted to the whole column.

Gravity intervenes in the abdominal cavity, although at the top the effect is diminished by the effects of thoracic suction. As we move inferiorly, the effect progressively increases; gravity becomes more noticeable as thoracic suction becomes less so.

The intervention of gravity is shown by differences in pressures in the abdominal cavity — as one moves inferiorly, pressure increases. The pressure in the abdominal cavity of a woman lying down has been measured as 8cm H_2O. With the subject standing, the pressure varies from 30cm H_2O in Douglas' pouch, to 8cm H_2O in the epigastrium and -5cm H_2O in the region under the diaphragm. This pressure may increase during contraction of the diaphragmatic or abdominal muscles when coughing, defecating or making a strenuous effort; it can transiently go higher than 100cm H_2O in the pelvis.

This confirms the relative weightlessness of the submesocolic region and explains certain phenomena such as the weakness of the support tissue of each viscera; the way a heavy viscera, such as the liver, is held in place; the frequency of ptosis of the stomach (which is under the effect of thoracic suction at the top and gravity at the bottom); and the frequency of diaphragmatic hernias, in which organs including the colon, or even the pancreas, can penetrate into the thoracic cavity.

The stacking up of the viscera (which is maintained by intracavitary pressure, turgor and the tonicity of the abdominal muscles) is as fragile as a house of cards and the smallest instability can cause serious disorders.

ABDOMINAL WALL

The muscular tone of the abdominal wall gives the visceral column its shape. Without it, neither intracavitary pressure, turgor nor the presence of the peritoneum would be enough to maintain the column. The intraperitoneal viscera (particularly those held least strongly) would sink into the internal iliac fossae and then overflow forward and laterally as happens in hara-kiri.

There are many possible causes of abdominal hypotonia, including everything from transient post partum hypotonia to the total paralysis caused by serious central nervous system disorders. Hypotonia causes the viscera to lose their cohesion and slide downward, pulling on the mesentery. Possible results include:

- ligamentous laxity (ptoses)
- inflammation (secretion of peritoneal fluid, causing an adhesion)
- muscular restrictions (viscerospasm)
- circulatory problems (venous stasis)
- transit problems (adhesion, occlusion)

Surgery is a frequent cause of intraperitoneal problems. In our daily practice, it is the most frequent cause of mechanical perturbations. This is not to deny that in many situations (e.g., acute appendicitis) surgery is the treatment of choice.

When a patient needs surgical intervention, he already suffers from inflammation and irritation of the peritoneum. When the peritoneum is irritated, the secretion of peritoneal fluid increases. This fluid becomes thicker and causes a process of adhesion in which the mesentery, folds of peritoneum, intestinal loops, etc. tend to stick together. These adhesions can sometimes have a beneficial effect as they tend to isolate the infected area from the rest of the serous membrane. But usually they disturb abdominal mobility.

Can the irritations caused by surgical intervention be any more pathogenic than that from the disease process? Definitely! Surgeons cut cleanly across planes of tissue, producing wide gaps that require stitching to be held together. Imagine the stresses this puts on the already damaged tissues and it is not difficult to understand how surgery produces particularly severe adhesions.

Introduction to Following Chapters

The abdominal viscera will be discussed separately in the chapters that follow. This separation is for didactic purposes only. The concept of osteopathy, as outlined by A.T. Still, is the affirmation of the united global functioning of the body. Keep this in mind as you read on. Visceral manipulation requires great precision, which comes only from a detailed knowledge of anatomy. We have included simple anatomical references in order to provide you with a framework within which to work. Consider these brief descriptions as only a reminder. Please study your anatomical texts intensively (especially those of you who are students) before beginning the fascinating work of visceral manipulation.

We have not devoted a chapter to the spleen and pancreas. We have not forgotten them, but these two organs have the peculiarity, under normal conditions, of being extremely difficult to palpate and impossible to dissociate from the surrounding structures. In this book, we have not included speculative theories but only our own observations. It is our opinion that the spleen and pancreas could benefit from an induction-based treatment, but certainly not from one based on direct manipulations. However, as we are not able to discern the actual motions of these organs, we felt it would be imprudent to describe them at this time.

Chapter Four: The Liver and Biliary System

Table of Contents

CHAPTER FOUR

The Liver and Biliary System

The liver is the largest digestive exocrine gland of the body, with a primarily metabolic and energetic role. It has a firm constituency, but is friable and fragile and protected by Glisson's capsule. It contains 500-1,000g of blood and weighs at least 2kg. Its internal temperature is often higher than that of the surrounding organs; certain hepatic veins can reach 40 degrees C. The osteopath must reserve special consideration for this organ because, as in any energy-based theory of medicine, the liver is an indispensable relay for treatment. Whenever there is a decrease in the motion of the liver, the patient feels tired (this is not true for the other organs, except the kidneys). Conversely, when the mobility or motility of the liver is improved, the patient will typically feel invigorated within 2 or 3 days.

Anatomy

The liver, which is oval in shape, is found in the right side of the diaphragmatic dome; it crosses the epigastrium to flow into the left hypochondrium. The largest part of it is on the right and its longitudinal axis runs transversely. Most of the axes of motion of the liver are determined by its connections with surrounding structures.

RELATIONSHIPS

Superiorly, the liver has a connection with the costodiaphragmatic cul-de-sac and heart via the diaphragm. These relationships were described in chapter 2. The diaphragmatic connection to the liver is primarily posterior. The anterior part of the liver is more inferior than the posterior part.

The important anteroinferior side of the liver can be reached directly when the patient is sitting down and leaning forward. This side faces inferiorly, posteriorly and toward the left; it is irregular and marked by two sagittal grooves (the right and left longitudinal grooves), which divide it into three zones. The left zone (left lobe of the liver) is concave and rests on the anterior face of the stomach, which leaves a large impression. The right zone corresponds to the external part of the right lobe and presents two faces. The anterior face, containing the colonic impression, is situated lateral to the gall bladder and rests on the hepatic flexure of the transverse colon. The posterior face contains the renal impression. The middle zone is the most important and is limited laterally by two longitudinal grooves which are united in the middle by a transverse groove. The three grooves form a structure (the hilum) shaped like a capital "H". The hilum separates the quadrate lobe (anterior) from the caudate lobe (posterior).

The anterior edge of the liver is oblique from bottom to top and from right to left. It runs along the edge of the floating ribs to the level of R9/10 and then comes in contact with the anterior abdominal wall at a distance of one finger's width under the xiphoid process. After this it passes underneath R6 or R7 on the left. It is normally palpable inferior to the ribs only in children. With normal inhalation, it moves downward 2cm, and with forced inhalation as much as 5cm.

Posteriorly, there is a marked indentation corresponding to the vertebral column. The upper border of the liver is at the level of the inferior vena cava, which makes a groove in it; this area is held together by dense fibrous tissue. The right part of the liver is directly connected to the diaphragm without intervention of the peritoneum.

GALL BLADDER AND BILE DUCT

The gall bladder is a very important organ in the practice of osteopathy. It is easy to treat directly and effectively; it is important for you to know this and not spend your time trying to work on it indirectly through the vertebral joints. The gall bladder is pear-shaped, roughly 10cm long by 4cm wide and has a capacity of 40-60ml. The largest part, the bottom, is directed anteroinferiorly. The body is oblique superiorly and posteriorly and is slanted to the left. The neck is to the left of the body and extends in an anteromedial direction. The gall bladder is surrounded by peritoneum and has a rich nerve supply. This is one reason why it can become very tender.

The bile duct follows the free edge of the lesser omentum and descends obliquely toward the left and slightly posteriorly. The common bile duct is a continuation of the bile duct and empties into the descending duodenum, at the level of the large caruncle, through the ampulla of Vater. It is 5cm long and has a diameter of 5-6mm. At the ampulla of Vater, its width is halved. It is oriented in an inferior and slightly posterior direction.

VISCERAL ARTICULATIONS

Ligaments

The coronary ligament is a strong band of fibrous tissue that extends along the posterior side of the liver, connecting it directly to the diaphragm without an intervening peritoneum. As with all the abdominal ligaments, it is composed of two layers at both extremities. Its insertions on the diaphragm (at the right anterior axillary line and the left midclavicular line) are called the right and left triangular ligaments respectively. The left triangular ligament is the most developed and is, along with intracavitary pressure, the primary element of maintenance and support for the liver.

The falciform (or suspensory) ligament is a thin peritoneal fold which connects the superior convex side of the liver to the diaphragm and anterior abdominal wall and divides the liver into a left and right lobe. It is triangular and contains remnants of the umbilical vein. Posteriorly, the falciform and coronary ligaments unite to form the horizontal and vertical components of a "T" *(Illustration 4-1)*. The falciform ligament does not play an important part in the support of the liver, but is useful for indicating the direction of liver movements.

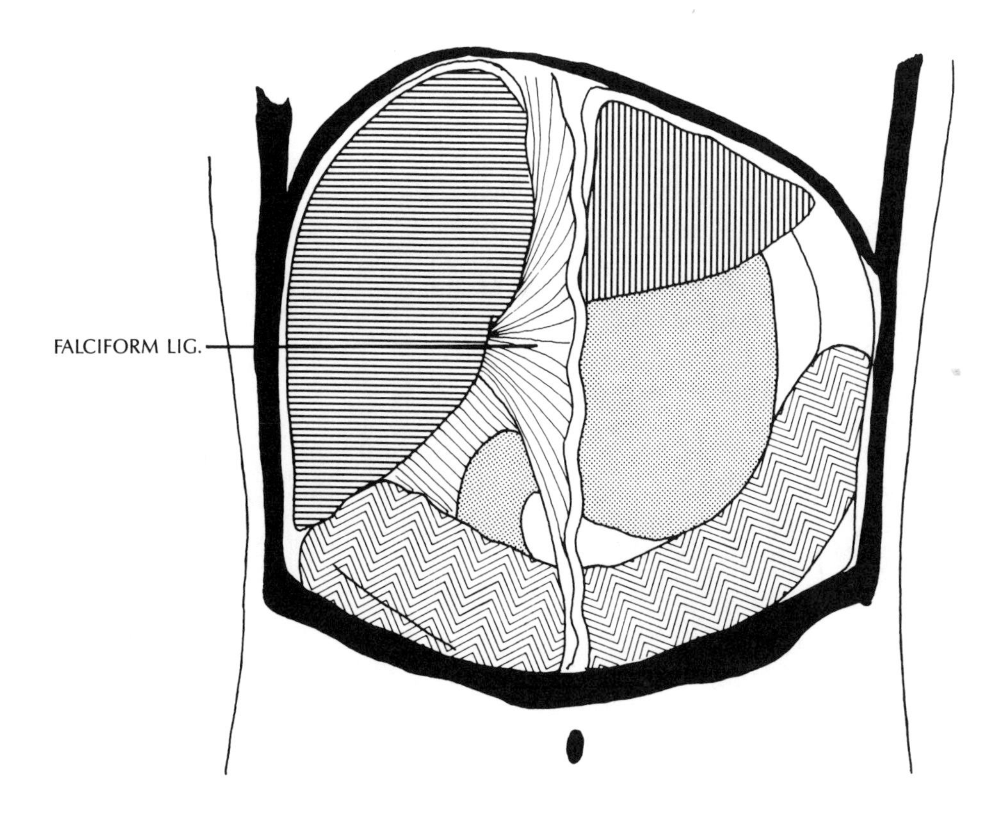

Illustration 4-1
Falciform Ligament

The hepatorenal ligament is another peritoneal fold which joins the liver to the kidney, mentioned here in order to emphasize the close contact between these two organs. As we will see in chapter 8, this connection is used to improve the mobility of the kidneys.

Other Articulations

The liver is closely connected via the hepatic vein to the inferior vena cava, which in turn adheres strongly to the diaphragm. The lesser omentum is a peritoneal fold uniting the liver to the esophagus, stomach and superior duodenum. It is covered by the liver and its anterior aspect faces left. Its gastroduodenal edge begins at the right side of the abdominal esophagus, follows along the lesser curvature of the stomach and ends at the superior duodenum. The lesser omentum comprises the hepatogastric ligament (which is furled like a sail) and the hepatoduodenal ligament (which is fixed to the descending duodenum, hepatic flexure of the colon and greater omentum). The hepatoduodenal ligament is relatively strong for an omentum. All of these connections have an important influence on the liver's motion.

Sliding Surfaces

The articular importance of the liver is emphasized by the impressions left on its inferior side by adjacent organs. Covering the entire inferior surface of the left lobe is the articular surface of the stomach, delimited to the right by the longitudinal groove of the falciform ligament. Further to the right and anteriorly is the colonic articular surface, which is lateral to the gall bladder and anterior to the renal impression *(Illustration 4-2)*. The superior and descending parts of the duodenum also leave impressions on the right lobe and the liver slides on the diaphragm. A hepatic restriction could affect any or all of these articulating organs.

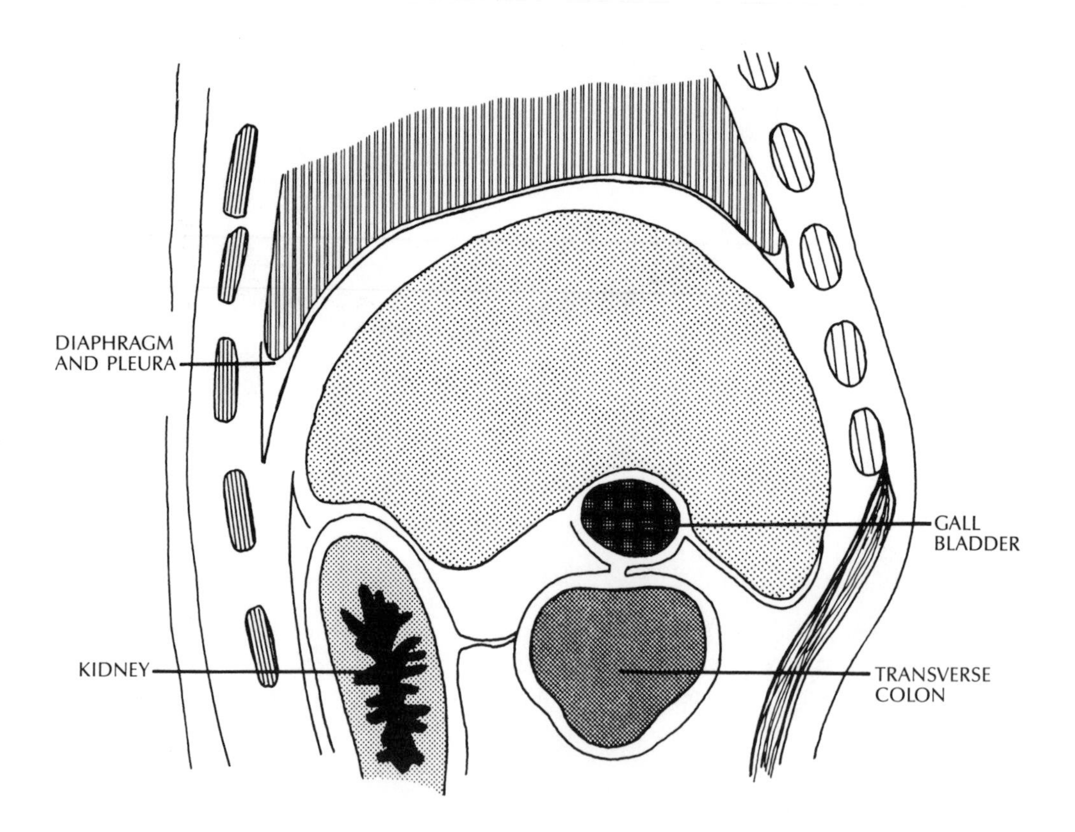

Illustration 4-2
Sliding Surfaces of the Liver

TOPOGRAPHICAL ANATOMY

The liver is 15-18cm in height at the right midclavicular line. Its superior aspect starts against the right lateral part of the diaphragm and ascends until it reaches the 5th intercostal space at the right midclavicular line. It ends, on the left, between the 5th and 6th intercostal spaces just medial to the left midclavicular line. It may extend below the xiphoid process, depending on the size of the thorax and the acuity of the costal angle *(Illustration 4-3)*. Posterosuperiorly, the liver is bounded by a line going through T8 or T9 and toward the lower part of R8 on the right. Posteroinferiorly, it is bounded by a line going from the upper part of T12 to R11 on the right. Its inferior limit is usually the inferior right edge of the rib cage.

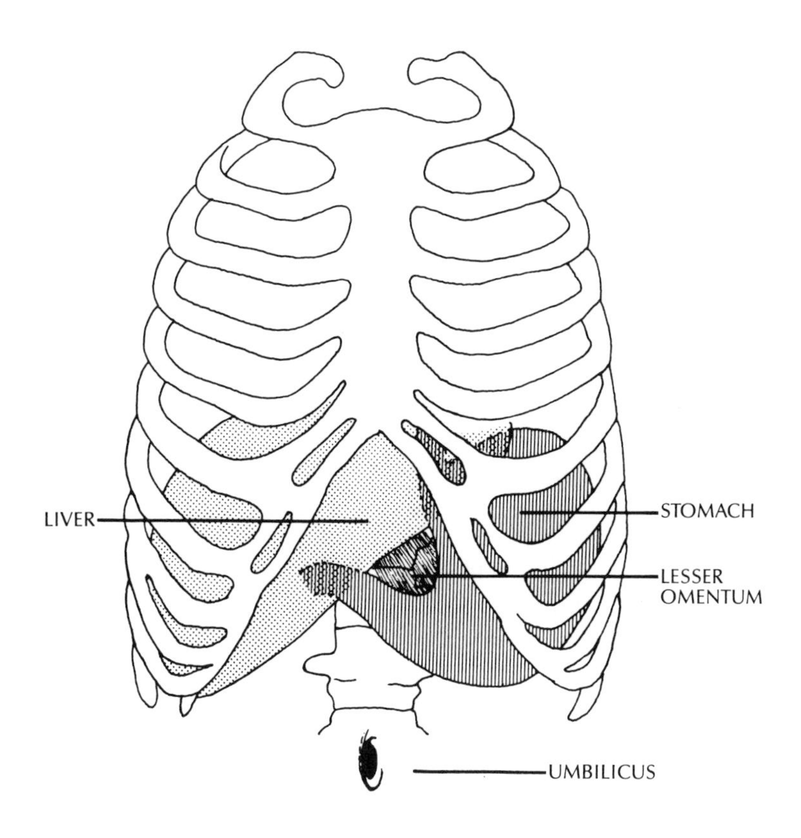

Illustration 4-3
Topographical Anatomy of the Liver

Depending on the subject, the gall bladder is located on a line connecting the umbilicus to either the right nipple or the right mid-clavicular point. The body of the gall bladder is located where this line crosses R10 on the right. The common bile duct is found slightly posteromedial to the line.

During palpation, from right to left, the fingers will encounter the transverse colon, right phrenicocolic ligament and, behind the colon, the anterior face of the right kidney, which is often a sensitive area. A little further to the left, and slightly anterior to the superior duodenum, you can palpate the body of the gall bladder at the point described above. Sensitivity here signifies an irritated and congested gall bladder. Further to the left is the longitudinal groove with the round ligament running along it, and finally the stomach *(Illustration 4-4).*

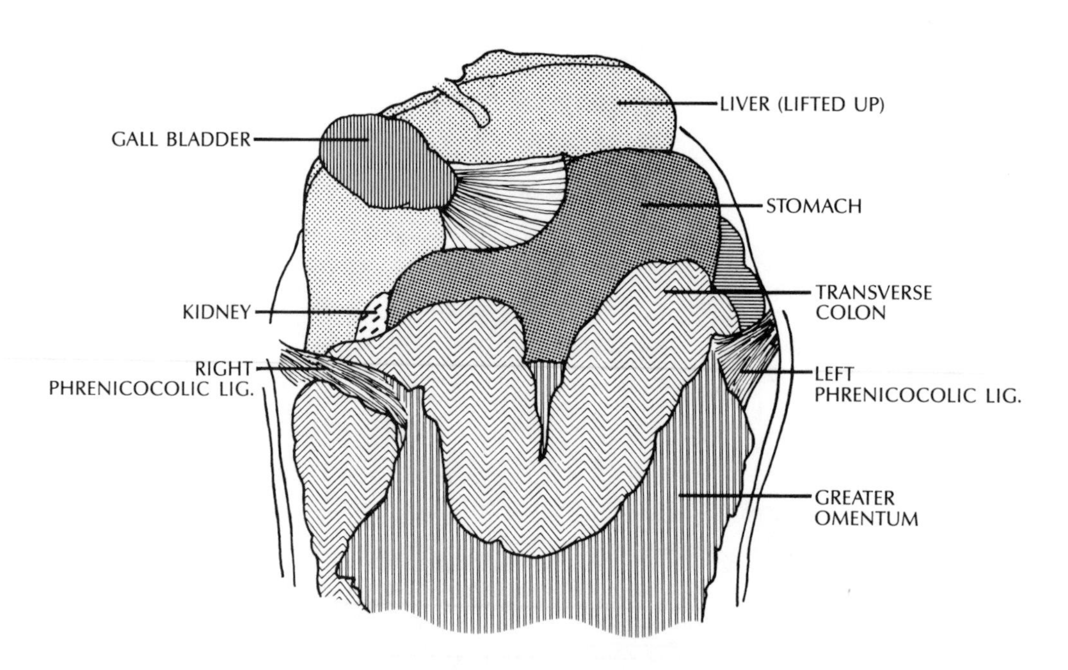

Illustration 4-4
Local Anatomy of the Right Upper Quadrant

The more the fingers move to the left, the more difficult it is to penetrate deeply. With overweight subjects, it is very difficult to carry out this approach; fortunately, such patients rarely need subcostal treatment — the liver is usually in good functional condition.

The lower biliary confluence (the union of the cystic and hepatic canals) is found in the lesser omentum at the lower edge of L1. The common bile duct then crosses the fixed part of the superior duodenum and goes right and downward to end in the ampulla of Vater at the posteroinferior aspect of the descending duodenum. At the level of L3, its anterior abdominal projection is 2cm above and to the right of the umbilicus. By beginning at the gall bladder and then moving the fingers slightly to the left, one can reach the upper part of the common bile duct across the superior duodenum. The sphincter is covered anteriorly by the transverse colon and the mesocolon; it is through these two organs that one can perform a biliary flushing technique which combines emptying of the gall bladder and manipulation of the common bile duct (pages 98-101).

Physiologic Motion

MOBILITY

The passive movement of the liver follows the movement of the diaphragm. Movements in the frontal plane are relatively easy to palpate compared to those in the sagittal and transverse planes. We shall discuss the movements of the liver during inhalation; those during exhalation are just the reverse.

In the frontal plane, the liver is so well attached to the stringy central part of the diaphragm that when the diaphragm descends, the liver follows. The diaphragmatic movement comes mostly from the back because the posterior diaphragm is a fleshy muscle, whereas the anterior part is more a thin musculoaponeurotic sheet. The diaphragmatic push is therefore a movement in an inferior and slightly anterior direction and the parts of the viscera adjacent to the diaphragm will move accordingly with inhalation.

The center of the diaphragm goes down less than the lateral parts because abdominal resistance is concentrated at this point. The lateral parts of the diaphragm will push the lateral part of the liver much further down and medially. Therefore, the whole liver moves downward and then rotates counterclockwise along the path of the falciform ligament, around an anteroposterior axis which passes through the left triangular ligament *(Illustration 4-5)*. During inhalation, all the diameters of the thorax increase and the lower ribs move upward and away from the median axis, in such a way that the thorax moves away from the liver, which in turn moves toward the median axis. You may be misled

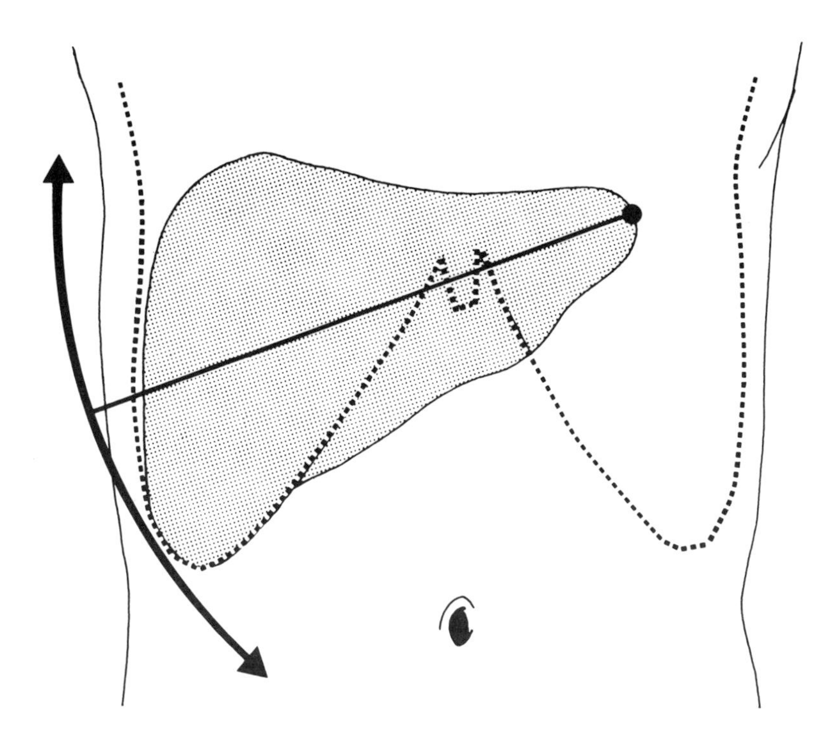

Illustration 4-5
Mobility and Motility of the Liver — Frontal Plane

by this rib movement into thinking that the liver is moving upward and outward with inhalation.

In the sagittal plane, the liver carries out a rotation at the end of inhalation: the anteroinferior side moves slightly posteroinferiorly to close the angle with the hepatic duct. One way to conceptualize this movement is to think of the liver as rolling forward. Remember that the diaphragm moves slightly anteriorly due to its concave shape. The posterior skeletal structures are less mobile than the anterior ones. The rotational axis

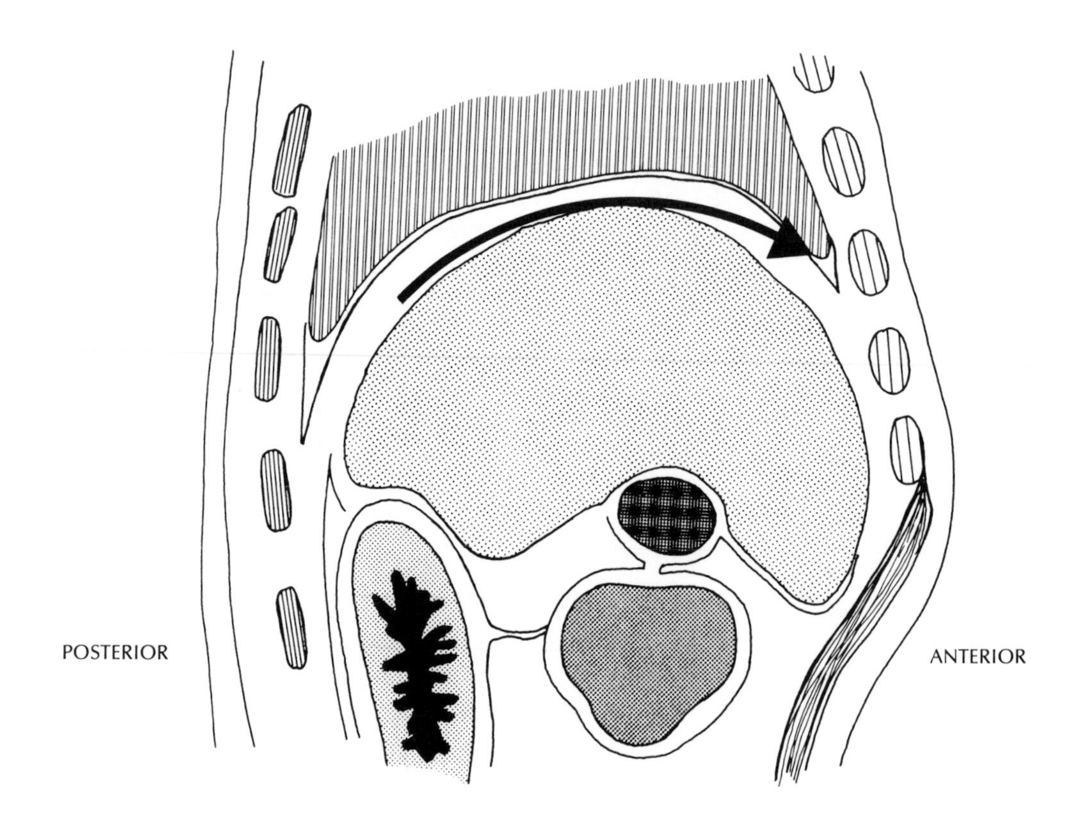

Illustration 4-6
Mobility and Motility of the Liver — Sagittal Plane

is on an oblique frontal plane that passes through the two triangular ligaments *(Illustration 4-6)*. This bi-triangular axis goes in a right to left and superior to inferior direction.

In the transverse plane, there is a slight rotation which is very difficult to appreciate. It seems that the lateral edge of the liver moves anteriorly and from right to left while the ribs move in the opposite direction. The axis of this counterclockwise (when viewed from above) rotation is vertical and goes through the inferior vena cava *(Illustration 4-7)*.

The result of the movements made on all three planes by the liver is complex but easy to appreciate; as the amplitude is quite large *(Illustration 4-8)*.

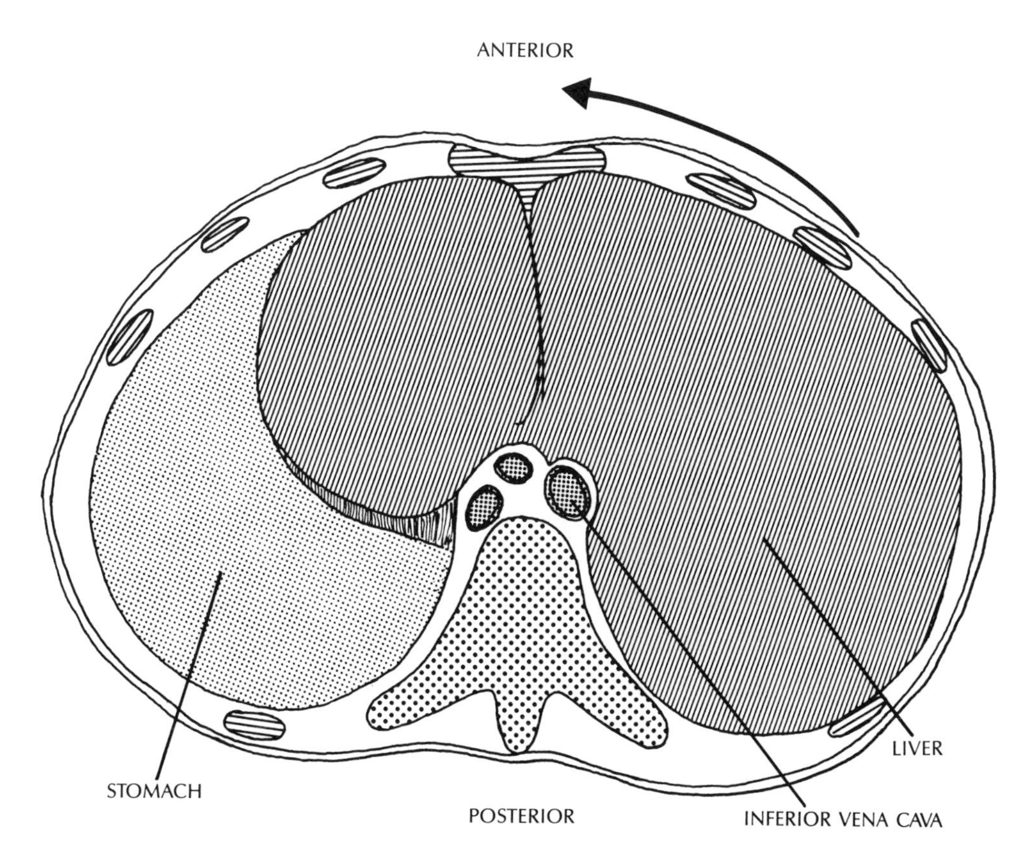

Illustration 4-7
Mobility and Motility of the Liver — Transverse Plane

MOTILITY

In almost every detail, the movements of motility are identical to those of mobility, but with a characteristically reduced amplitude and slower rhythm. The following motions take place during expir (and the opposite motions during inspir). These motions are comparable to the inhalational movements of mobility; please refer again to Illustrations 4-5 through 4-8.

In the frontal plane, the liver makes a counterclockwise motion around an anteroposterior axis which passes through the left triangular ligament. Along with the motion of the lung, this is one of the most important examples of visceral motility that exists. It is relatively easy to palpate and, because of the liver's many connections, can be used as an indicator for the general condition of the viscera.

In the sagittal plane, the liver rotates or "rolls" forward around a bi-triangular axis in a fashion similar to that of mobility. This motion is minimal and difficult to appreciate.

In the transverse plane, the lateral edge of the liver moves in a counterclockwise rotation (when viewed from above) from back to front and from right to left. Although the amplitude is small, it is very important to carry out this movement in induction techniques, as the ability to re-establish it is an important criterion for success.

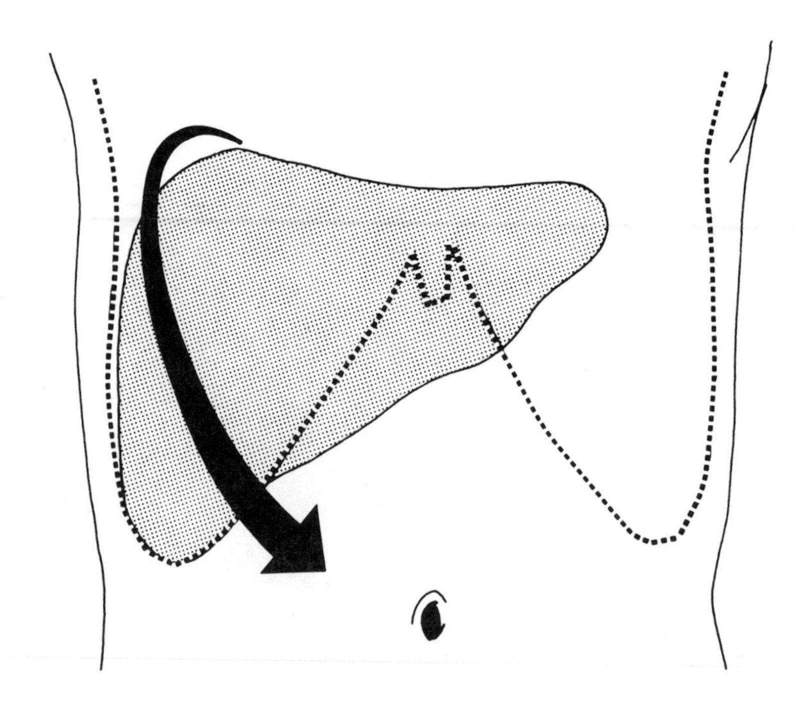

Illustration 4-8
Mobility and Motility of the Liver — Resultant Motion

Indications for Visceral Evaluation

Because of the importance and wide-ranging effects of its functions, the liver plays a central role in visceral manipulation. Manipulation of the liver has a double role: to optimize its metabolic activities and increase the flow of blood, lymph and bile through it. It is rare that we do not have to restart function of the liver in women, due to its role in metabolism of hormones (including the estrogens). The liver has an increase in workload after ovulation and manipulation performed during the first part of the cycle will be much more effective than that done after ovulation.

In men, disorders of the liver are rarer, with the exception of viral infections or toxicity from alcohol or other chemicals. The indications for manipulation of the liver in men are more commonly for disorders of the excretory paths (e.g., spasms of the bile duct, cholecystitis, cholelithiasis, high cholesterol levels and gastric bile reflux).

Gastric bile reflux is an indication for manipulation of the bile ducts. Gastritis and duodenal ulcers are made worse by the alkalization of the stomach walls by the bile salts, as the gastric mucosa prefers acids to salt bases and the presence of the latter disturb gastric secretions. Commonly observed, but difficult to explain, are connections between the liver and the right eye and frontal area, and between the gall bladder and the left eye, face and neck. These cases often involve a restriction of C4/5 with very slight motion (in contrast to the total restrictions of this area that occur after whiplash injuries). Very often, problems with these areas (including craniosacral restrictions) are related to liver or gall bladder dysfunction. Interestingly, there are often sensations appreciated by the patient in these areas during manipulation of the liver or gall bladder.

Another important indication for manipulation of the liver is nervous depression. The relationship between depression and the liver is well known in traditional Oriental medicine. In this tradition, as understood in modern France, the liver deficiency is derived from a lack of energy in the brain, particularly from the right frontal areas; this in turn depletes the brain's energy. The finding of biliary pigments in cerebral tissue would tend to support this concept.

We have found, especially in children, positive short- and long-term results from liver manipulations in patients with chronic or recurrent sinusitis and/or bronchitis. While we have no hard proof at present, we believe that the efficacy of these techniques in treating these problems points to a beneficial influence on the immune system.

In conclusion, the primary indications for hepatic manipulation include:

- decrease in hepatic metabolism
- biliary stasis
- nervous depression
- decrease in immune responses

Another possible indication for manipulation of the liver is musculoskeletal dysfunction of the thoracic inlet, particularly periarthritis of the right shoulder. There is a simple way to determine if the liver is involved in this condition. Test the range of motion of the joint and then inhibit the liver (by gently pressing on the right tenth intercostal space at the midclavicular line) and repeat the motion of the shoulder with the greatest restriction. If the liver is involved, you will see an immediate improvement of 30-40 degrees of motion.

Indications for manipulation of the gall bladder include the classic right upper quadrant and/or right scapular pain that is precipitated by greasy food. Gall bladder dysfunction may also cause pain on exhalation, as well as problems with the left eye, left frontal area or left neck. Often these problems become worse after surgery as the common bile duct becomes irritated and edematous, resulting in stasis. This duct (along with the pylorus) should always be treated after surgery.

Techniques to improve mobility are used when there are important restrictions of the liver such as sequelae of a viral, parasitic or bacterial illness. In these cases the liver, affected in its entirety, fixes itself in position. In the same way, the sequelae of pleuropulmonary invasive illnesses can restrict the mobility of the liver. In the case of diaphragmatic adhesions accompanied by characteristic crepitations, the liver too closely follows the movements of the diaphragm. Movements of the liver in a frontal plane remain within normal limits, whereas movements on the other planes are affected and motility becomes virtually nonexistent. Mobilization techniques can lead to the release of these adhesions. Once again, we must emphasize that evaluations must not stop at the level of grossly apparent movement; a normal cholecystography is not sufficient proof of a properly functioning liver.

Evaluation

The evaluation will depend upon your capacities and background. For a list of appropriate questions, please consult the proper medical texts. It is necessary to gain a knowledge of past infections or a history of problems from vaccinations, as these often impair the function of the liver.

Examination of the liver should include percussion. In case of any doubt, do not hesitate to use the resources of the modern laboratory, e.g., biochemical and imaging technologies. *Using alternative or natural medicine should not mean rejecting the techniques of conventional medicine.*

MOBILITY TESTS

All the mobility tests described here and in the following chapters require some relatively deep palpation of the abdomen. In our daily life, it is very rare to have someone place an anteroposteriorly directed pressure on our abdomen. This, combined with a feeling that the abdomen is a relatively private place, invokes a reticence and some mental defenses against abdominal palpation in most patients. It is therefore important to conduct this part of the examination (and treatment) with gentleness and care.

We examine patients for problems of liver mobility directly, via a subcostal route. The patient sits bent forward in order to release the tension of the abdominal muscles. This makes it easier to penetrate deeply with the fingers without causing discomfort.

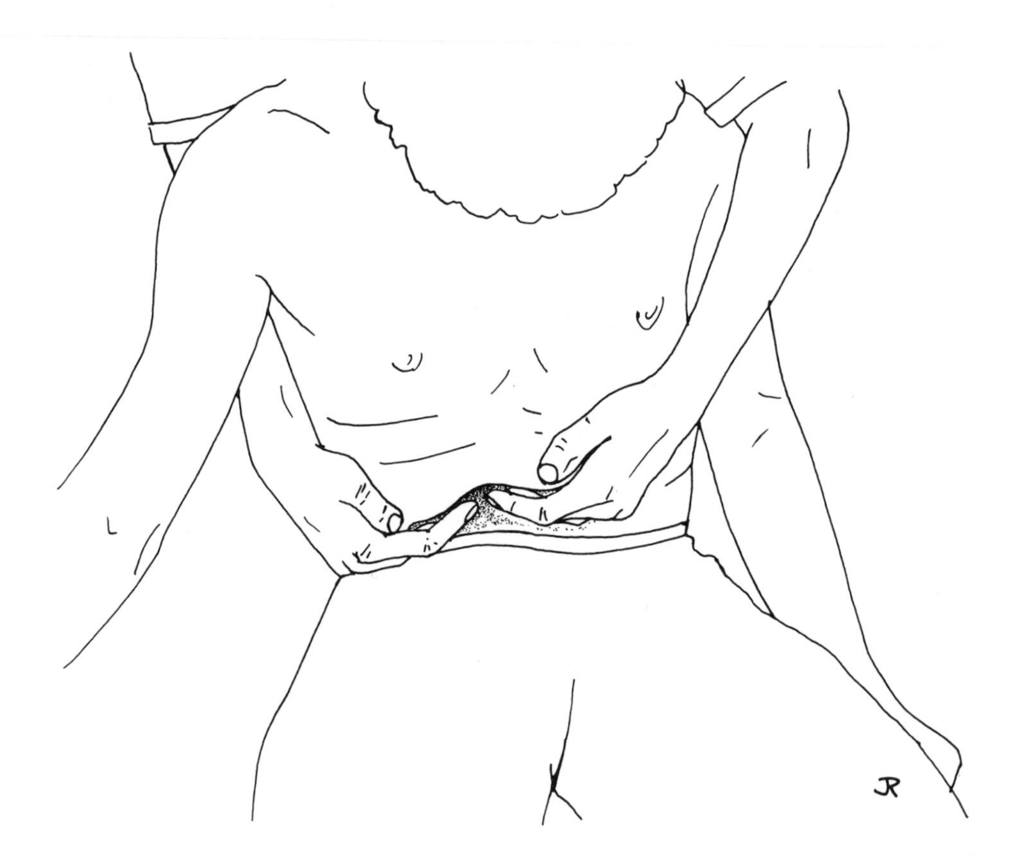

Illustration 4-9
Direct Subcostal Approach to the Liver

In the seated position, gravity brings the liver downward. Place your fingers with their ulnar edges just under the patient's lower ribs and parallel to them. Go to the lateral edge of the rectus abdominis muscle in order to avoid the normally tender gall bladder. The ulnar edges are used because they cause much less discomfort than the finger tips *(Illustration 4-9)*. Gradually increase the pressure, at first in a posterior direction. When your fingers can no longer move straight posteriorly, change to a posterosuperior direction. The more forward bent the patient, the deeper your fingers can go.

Direct Tests

When you push the central part of the liver posterosuperiorly as described above, it should move by 1-2cm. Of all mobilizations, that of the liver is the most objective because no other solid organ directly interferes with or gets in the way of its movement. Depending on the location and direction of the pressure, you can test several structures of the liver. Many combinations and variations of the tests described here are possible.

Pushing the liver superiorly tests the level of restriction in the inferior structures. For example, excessive resistance to this movement could reflect a problem with the hepatorenal ligament. Appreciating the amplitude and speed of the liver's return inferiorly, as you release the pressure, tests for restrictions in the superior structures. When there is a restriction, the return will be slower than normal and of a shorter distance. This test can be made more specific by modifying the direction of the push and the position of the patient. For example, sidebending the patient to the left and directing the posterosuperior push toward the right shoulder will allow a finer appreciation of any restrictions in the right triangular ligament.

The left triangular ligament can sometimes be evaluated fairly directly. Place your fingers just to the left of the xiphoid process. Sidebend and rotate the patient to the left as you come under and just posterior to the ribs. From this position, a pressure directed towards the left triangular ligament from your fingers should allow you to evaluate that structure.

To fully appreciate the sagittal or transverse motions of the liver, it is necessary to push in a primarily posterior direction. If you push your fingers directly toward the back, this will lift the posterior aspect of the liver and the organ will fall forward. A delay or decreased distance in the return of the liver in this situation is related to the function of the anterior part of the coronary ligament.

Indirect Tests

The elasticity of the lower ribs may be utilized for indirect evaluation of liver mobility. These tests may be done in a supine, left lateral decubitus or seated position. The ribs are compressed as you check for abnormal resistance. A feeling of normal costal elasticity but a resistance underneath them indicates some problem of liver mobility. Depending on the direction of costal compression, the areas of liver restriction can be localized. However, these techniques are nonspecific because there are other possible reasons for abnormal rib movement, primarily musculoskeletal dysfunctions. Before testing the right side, you should test the left side for the purpose of comparison.

With the patient in the supine position, place the palm of your hand on the inferior anterolateral aspect of the right thorax and compress the ribs, pushing them anteriorly, inferiorly and medially. This technique should be painless and progressive. The shape of the thorax should be easily deformed (think of all the variation due to sex, body type and age). During the movement, place your other hand under the costal edge. Normally, except in children, the liver does not spill over the costal edge.

With the patient in the left lateral decubitus position, push the right costal mass anteriorly and inferiorly, using the same methods as in the supine position.

With the patient in the seated position, stand to his left. Surround the lower right ribs with your joined hands and compress the thorax, bringing it towards you, while at the same time rotating it first to the left and then to the right. The articulation tests of the costovertebral and costochondral joints corresponding to the liver can be done in association with these tests.

MOTILITY TESTS

With the patient in the supine position, place your right hand on the hepatic region, fingertips as close to the left triangular ligament as possible, and palm on the lateral part of R9-11 above the right lateral aspect of the liver. Your hand should follow the convexity of the rib cage; in order to concentrate better, it is a good idea to place your left hand on top of your right. You must be passive in order to be objective. If the motion is difficult to appreciate, or in order to free your mind, try visualizing the anatomy of the liver —

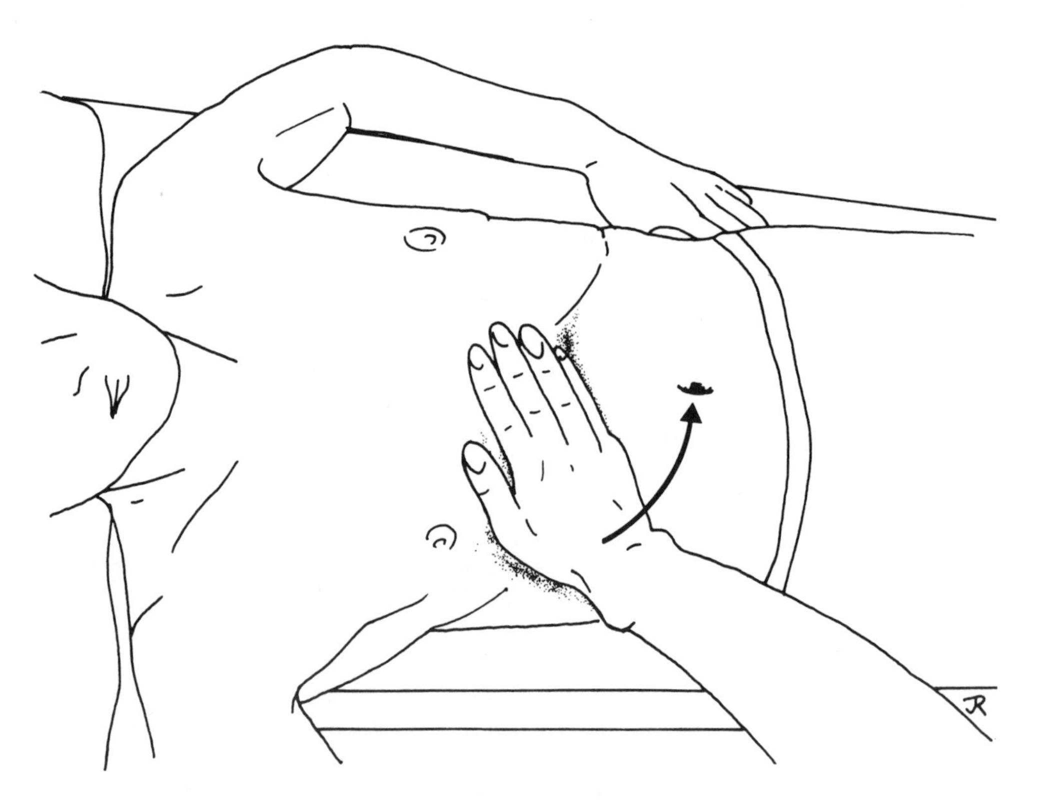

Illustration 4-10
Motility Test of the Liver — Frontal Plane

this is a good way of becoming precise and of enhancing your perception. For the beginner, it is usually easier to palpate the motion during the expir phase.

In the frontal plane, during expir, your hand should rotate from right to left in a counterclockwise motion around an anteroposterior axis that passes just distal to the third knuckle of the right hand. In this motion the palm of the hand moves toward the umbilicus *(Illustration 4-10)*.

In the sagittal plane, during expir, the superior part of your hand rotates anteriorly and inferiorly around a transverse axis through the middle of the hand. The inferior part of your hand rotates in a similar manner and therefore seems to press harder against the body *(Illustration 4-11)*.

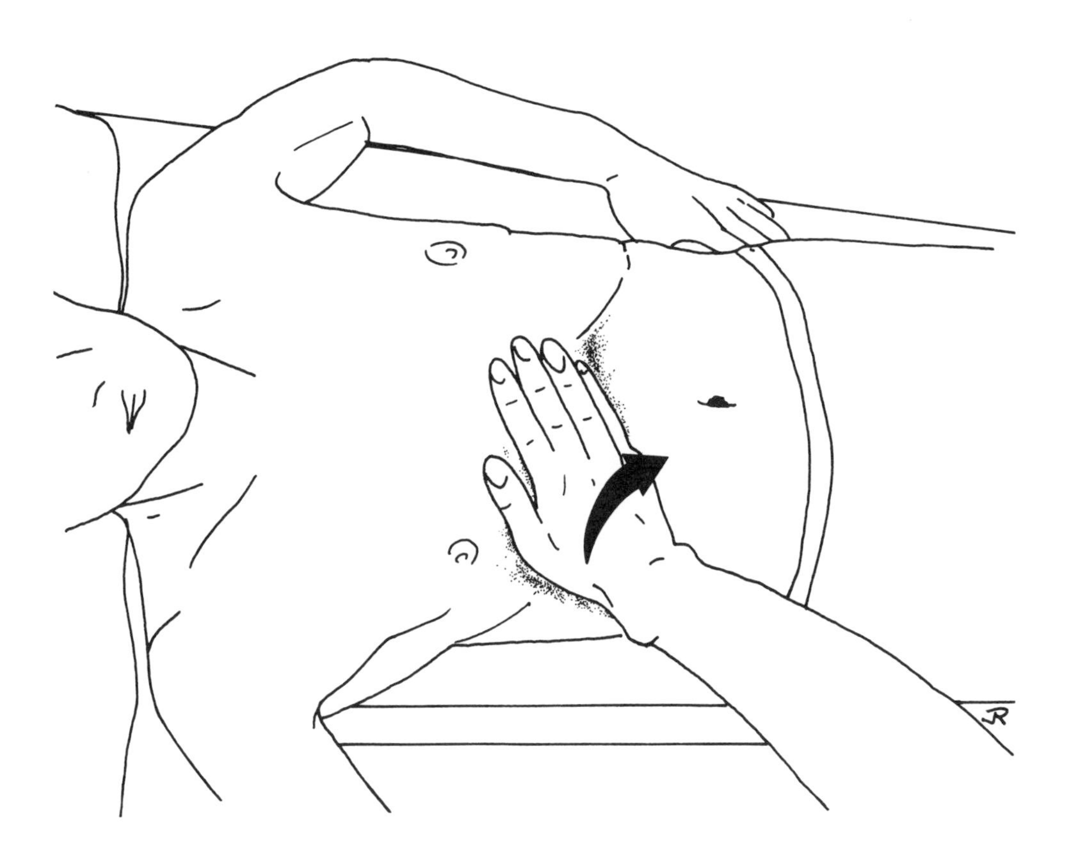

Illustration 4-11
Motility Test of the Liver — Sagittal Plane

In the transverse plane, during expir, your hand rotates to the left around a vertical axis that passes through the knuckles of the sensing hand. This seems to bring the palm away from the body while the fingers press in harder *(Illustration 4-12)*. After you become comfortable testing these different planes one at a time, you can test them all simultaneously for a more complete appreciation of the motility of the liver.

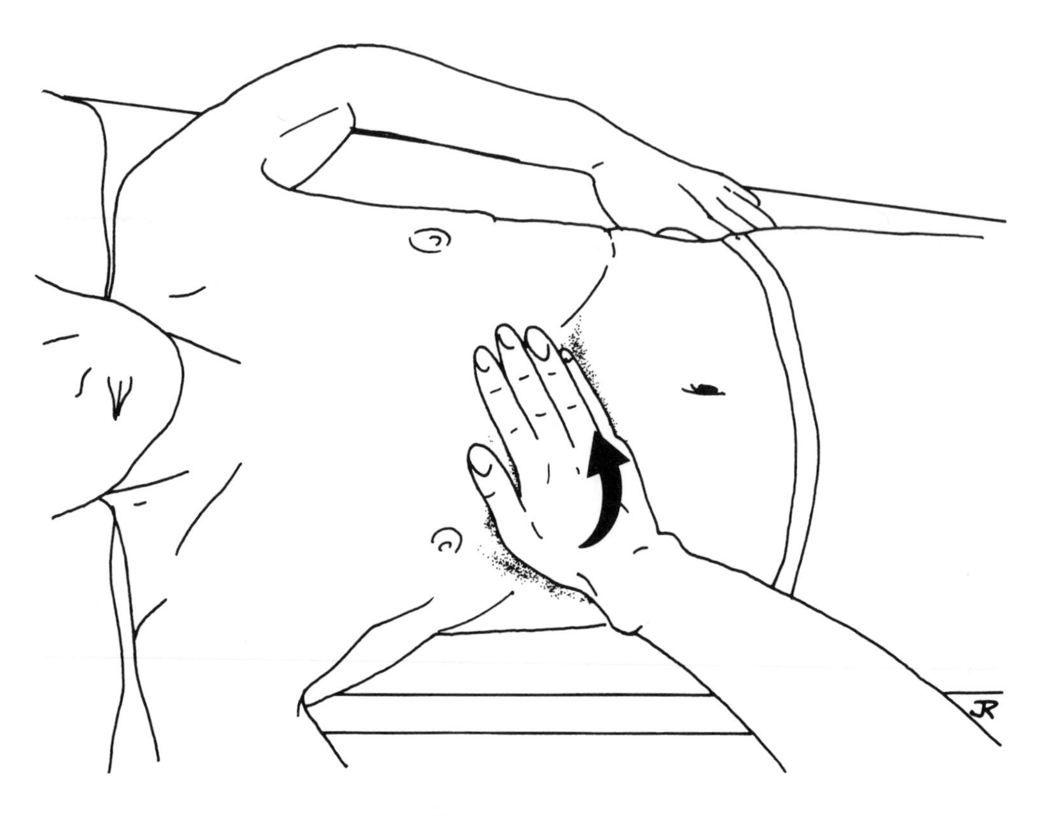

Illustration 4-12
Motility Test of the Liver — Transverse Plane

Restrictions

There are many types of restrictions involving the liver; we will mention some common ones. With hepatitis the liver will feel tight, while in cases of chronic obstructive lung disease it will feel heavy. Occasionally a total fixation of the liver can create a fibrosis of its supporting ligaments; this is common in cases of cirrhosis. During direct mobilization techniques you can feel these restrictions let go little by little, occasionally with a crackling sound. It is a marvelous sensation and reinforces one's appreciation of the efficacy of this technique. Posterior restrictions are common, perhaps because of the pleuropulmonary relationship. Every pleural problem directly affects the mechanics of the liver through restriction of its connections with the diaphragm. For this reason, it is imperative to check the motions of the liver in the aftermath of pulmonary diseases, and vice versa.

Manipulations

Manipulations to improve mobility should usually precede those to treat problems of motility. Frequently, a dysfunction in the liver will not involve the gall bladder, and vice versa. If both organs are involved, treat the gall bladder first.

DIRECT TECHNIQUES

Seated Position

As noted above, mobilization of the liver with the patient in a seated position enables you to release restrictions as far toward the back as possible. The direct technique is just a rhythmic, slow (approximately 10 per minute) repetition of the diagnostic technique; it involves evaluating the quality of motion by lifting the liver posterosuperiorly in a gentle and progressive manner by 1-2cm and then releasing it. The motion is inversely proportional to the amount of restriction in the coronary ligament; the more severe the restriction, the less the excursion *(Illustration 4-13)*. As the technique is repeated, the ligament will release (sometimes with a popping sound). Five to six repetitions should suffice. It is better to begin with the external part of the liver (the most mobile) at the level of the colonic impression in order to first release the tension of the hepatic flexure of the colon via the right phrenicocolic ligament. Then move toward the falciform and stomach areas where, to treat the area of the left triangular ligament, the fingers should be pushed superiorly and laterally. To work on the entire liver at once, put one hand medially and the other laterally. As you move your fingers to work on different areas, adjust the position of the patient to focus your effects on those areas.

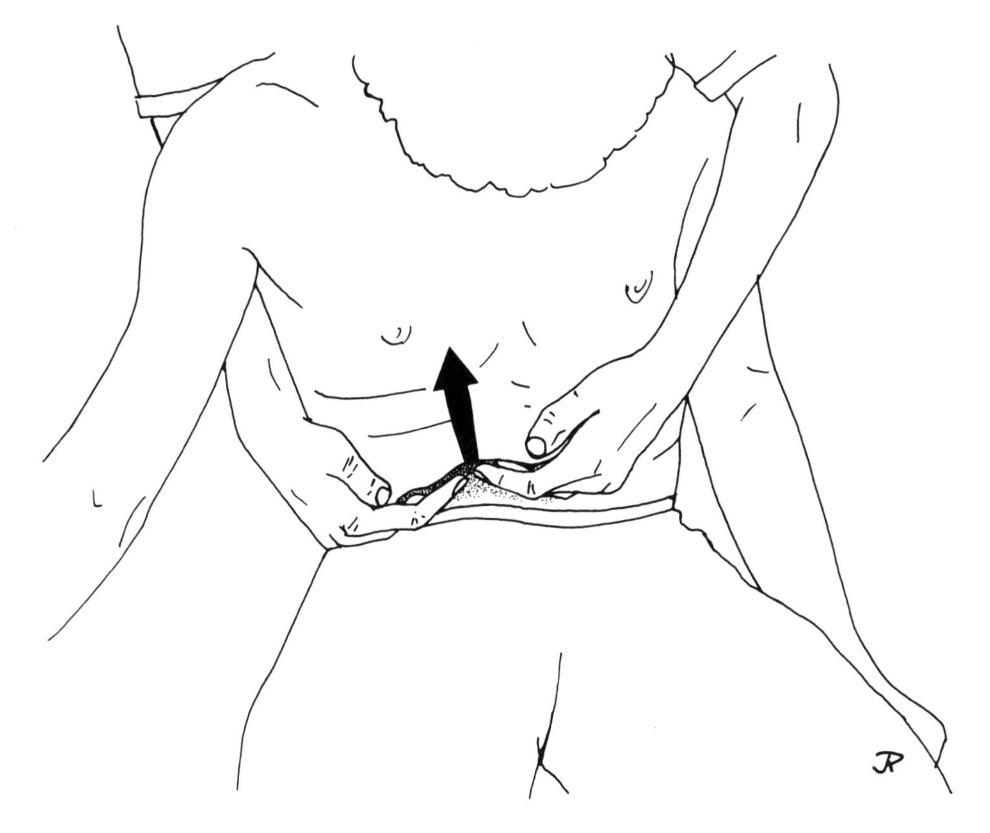

Illustration 4-13
Direct Manipulation of the Liver — Seated Position

This is very similar to manipulations for the stomach (see chapter 5). Of course, all the organs and structures attached to the liver are affected by this technique. When the tension of the anterior organs is released, the liver must be pushed up; you can then direct your fingers more toward the back to try to reach the anterosuperior part of the kidney, if desired.

Left Lateral Decubitus Position

Place yourself behind the patient and put the palms of your hands on the anterolateral extremities of R7-9, with the pads of the fingers subcostal. Push the ribs anteriorly, inferiorly and medially (toward the umbilicus) and then in the opposite direction, in a slow back-and-forth cycle of around 10 per minute with steadily increasing amplitude *(Illustration 4-14)*. It is not necessary to synchronize the movements with respiration. You can adjust the technique so that it focuses on motion in one particular plane. For frontal plane motion, push the ribs toward the umbilicus with more movement of the lateral aspects. For motion in the transverse plane, roll the ribs toward the midline. For sagittal plane motion, press your thumbs together on the posterior axillary line and use your fingers to rotate the ribs forward. Recoil (pushing medially and anteriorly in all planes and then releasing suddenly) can also be performed here.

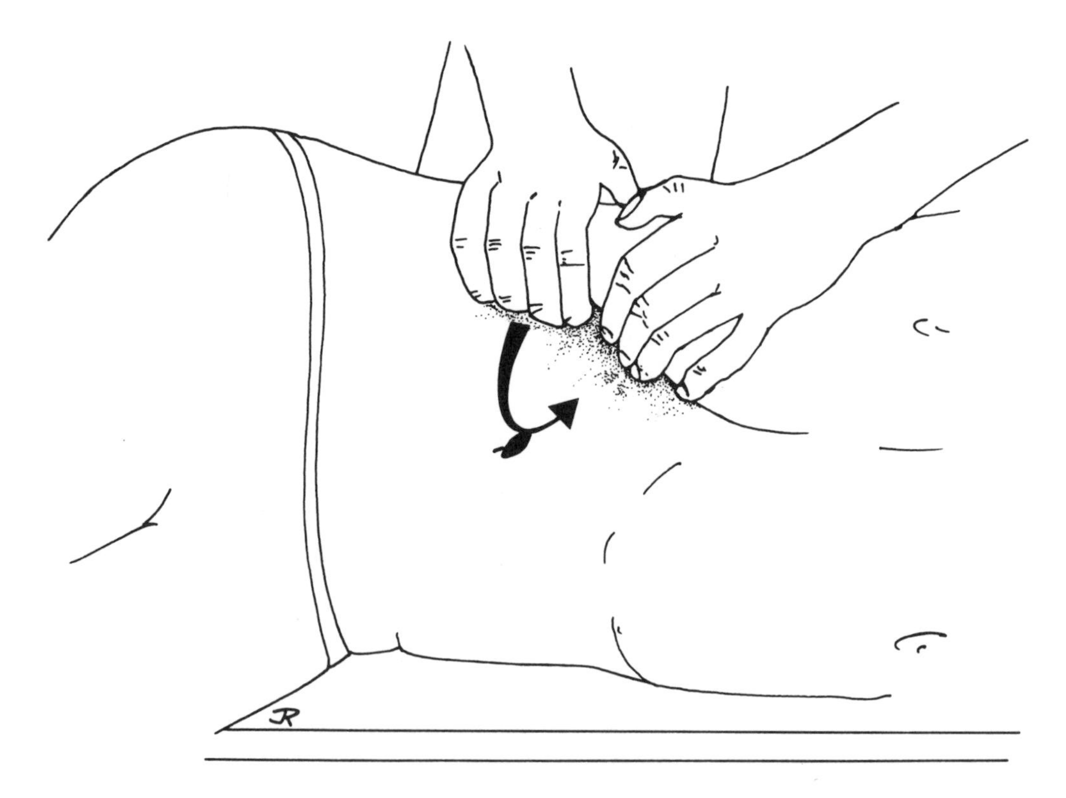

Illustration 4-14
Direct Manipulation of the Liver —
Left Lateral Decubitus Position

This technique moves the ribs, pleura, peritoneum and liver and is a good general mobilization which is tiring for neither you nor the patient. It is especially useful in patients who have had hepatitis. In general, however, techniques performed through the abdominal wall are more effective. Techniques which work through the ribs are used primarily when the abdomen is too tense, or as an adjunct.

Emptying of the Gall Bladder

Place the patient in a seated position and stand behind him with your fingers subcostal, just lateral to the gall bladder, as here the abdomen is more easily depressed. Flatten your fingers well against the inferior side of the liver and be careful not to confuse the gall bladder with the first part of the duodenum or the transverse colon. The superior duodenum is not sensitive to palpation, whereas the gall bladder often is. For this reason, first look for tender points in this area. You can easily reach the bottom of the gall bladder, which has a relatively anterior position. To force it to contract (and expel the bile or sediments which could be in it), push your fingers rhythmically with a moderate amount of force in short (2-4cm) strokes superiorly, posteriorly and medially along the long axis of the gall bladder. As you continue, the gall bladder will become less sensitive. This is the first sign of an effective treatment. If the gall bladder is tender it is

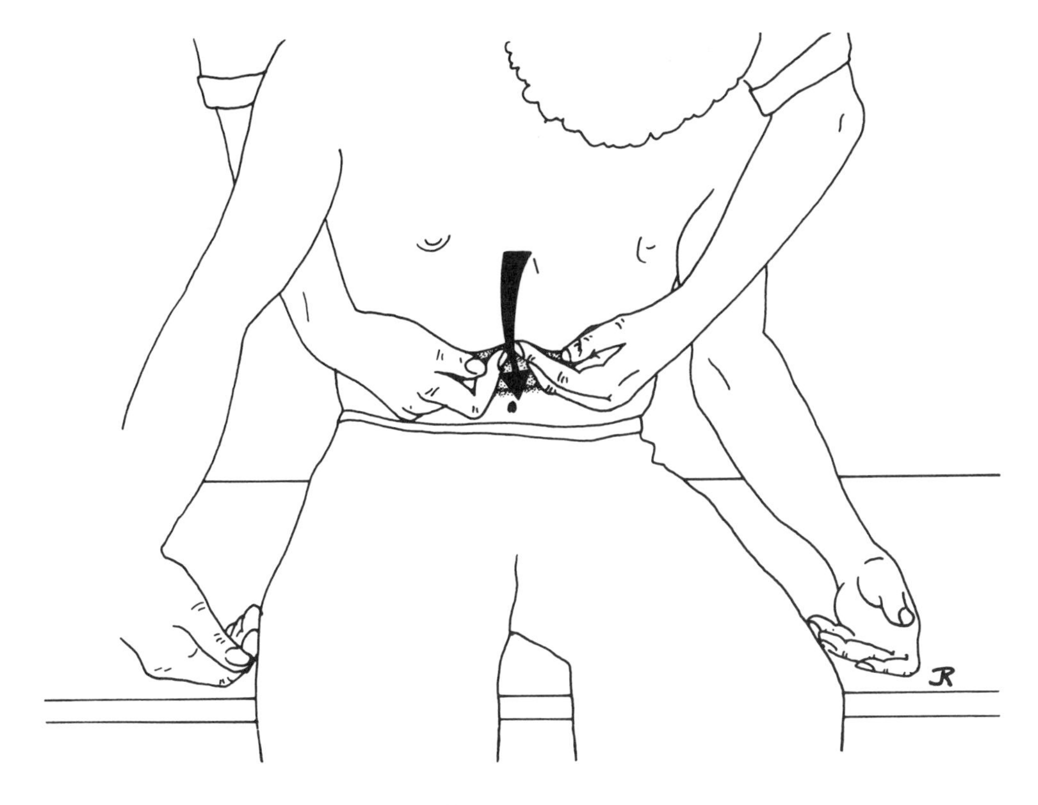

Illustration 4-15
Manipulation of the Common Bile Duct — Seated Position

important to do this technique gently at first (4-8 times) until the tenderness has dissipated and then increase your force. At this point, change the direction of your force to inferior, posterior and medial to release the neck of the gall bladder. Repeat this motion until there is a release.

In cases of cholelithiasis, there are risks of microhemorrhages or parietal inflammation if you press too hard. If the pain is too sharp, release the pressure and use your fingers gently and carefully. In the case of cholecystitis, carry out an antispasmodic movement like those done for the pylorus of Oddi's sphincter (see chapter 6). After the spasm decreases, you can increase your pressure or continue with the technique described previously.

Manipulation of the Common Bile Duct

Following the emptying of the gall bladder, flow in the bile ducts needs encouragement. The common bile duct (a smooth canal with only one sphincter at its opening into the duodenum) has fibromuscular walls and is able to reduce the size of its lumen. Its optimal function therefore depends on appropriate tone.

Use the seated position (page 97) which increases the patient's kyphosis and enables you to reach the deeper areas. The common bile duct is found very far back,

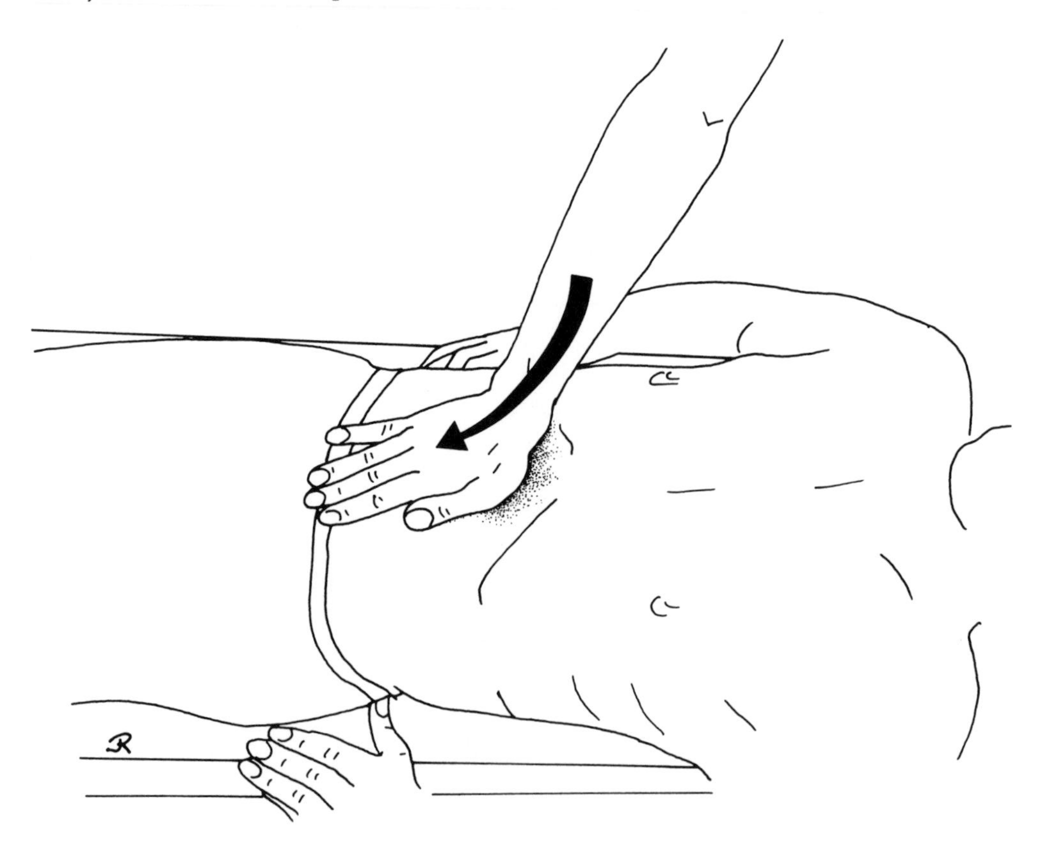

Illustration 4-16
Manipulation of the Common Bile Duct — Supine Position

posterior to the duodenum, at the T11/12 junction near the vena cava and epiploic foramen. Lower down, it is posterior to the pancreas. You therefore have to work on the most accessible part, which is that posterior to the duodenum. The anterior projection of the duct behind the duodenum is on the midclavicular-umbilical line, two fingers' width under the edge of the rib cage. Sometimes sidebending the patient to the left makes it easier to work on the duct. Push your fingers or thumbs toward the back and, when you have gone as far as possible, gently stretch your thumbs downward toward the umbilicus and release the pressure very slowly. Repeat this procedure slowly and rhythmically. When you have become comfortable performing this movement, you can add an oblique lateral to medial direction to it which results in an arc *(Illustration 4-15)*. This movement often provokes a characteristic discharge sound from the body. For best results, continue downward bit by bit toward the sphincter of Oddi as you feel each part of the duct relax.

For a patient in the supine position with knees flexed, stand at the patient's right. Press your thumbs or palm at the point on the abdomen described above (overlying the common bile duct and the duodenum), first toward the back, then toward the feet and slightly to the left *(Illustration 4-16)*. This movement is more difficult to perform and demands greater precision than that performed in the seated position, as the aid of gravity is not enlisted.

COMBINED TECHNIQUES

Seated Position

This is a variation on the direct technique in the seated position. As before, press your fingers against the lower border of the liver, preferably with one hand placed medially and the other laterally in order to appreciate the whole liver. Next, lean on the patient's shoulders and maneuver the body into the position where maximal stretching occurs for the area you wish to focus on. The direct aspect (short lever) is the hand or hands on the abdomen. The indirect aspect (long lever) is the use of a distant part, the shoulders, for positioning to influence the target area.

Using this positioning, the torso can be "played with" to find the correct position or series of positions to release restrictions of the liver. The patient, who feels safe and relaxed under the control of the practitioner, lets go and moves freely. The thoracic movement allows a precise multidirectional treatment to take place effectively and without pain. For example, left rotation and right rotation of the torso cause your fingers to be positioned more posteriorly or anteriorly, respectively. If the patient leans forward, you can reach the posteroinferior region more easily; if he bends to the side, you can reach the opposite side more easily.

Manipulation of the left triangular ligament is more difficult to perform. The patient must be bent forward in order to bring your fingers as far posteriorly as possible and then sidebent to the right while you push your fingers as far as possible toward the left. Maintain this pressure and the position of your hands while the patient sidebends to the left, which brings the left extremity of the liver inferiorly. This makes it easier to raise this part of the liver with your hands. Utilizing these movements of the torso, many combinations of techniques are possible *(Illustration 4-17)*. This is an extremely useful technique and can be adapted for use on the gall bladder or common bile duct.

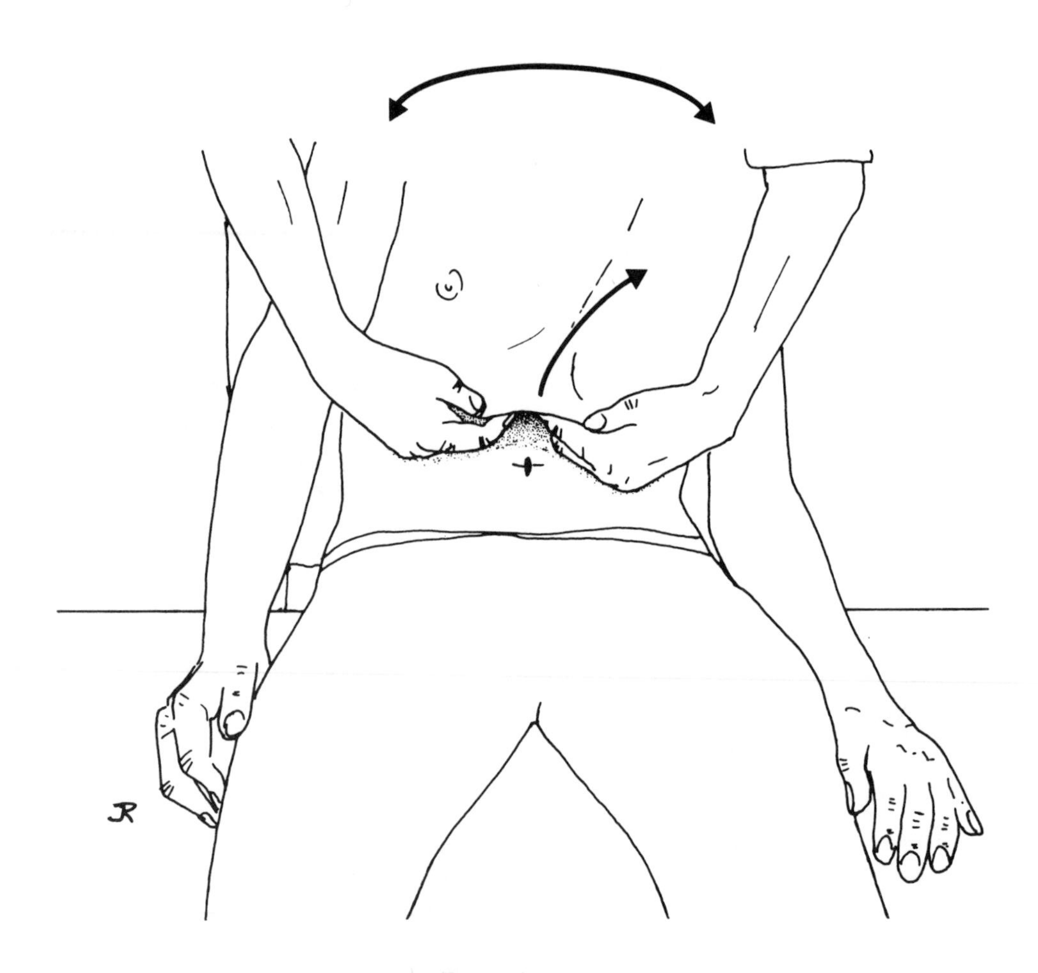

Illustration 4-17
Combined Manipulation for Release of Left Triangular Ligament — Seated Position

Another helpful combined technique stretches the common bile duct. It is usually applied after the direct release of the duct and before treating the sphincter of Oddi. In the kyphosed seated position, fix the duct just superior to the sphincter by a strong posterior pressure. While maintaining this pressure, extend and rotate the patient's torso to the right, which increases the distance between the fixed point and the diaphragm/liver. This effectively stretches the entire duct. Repeat this technique 3-5 times until you sense a release. The movement can also be done via the patient's elbows, with his hands clasped behind his neck.

Supine Position

A similar technique can be performed with the patient in the supine position. The short arm lever is as described above, but the long arm lever is accomplished by utilizing the lower extremities. With both the hips and knees flexed, the lower extremities can be rotated and/or sidebent to get to the position necessary for greatest effect *(Illustration 4-18)*. If you perform rotation and sidebending to the right, there is a compression effect on the liver. Sometimes this can lead to a point of balanced tension that will result in a very significant release. If the movements are performed to the left, then the result is a stretching of the structures concerned.

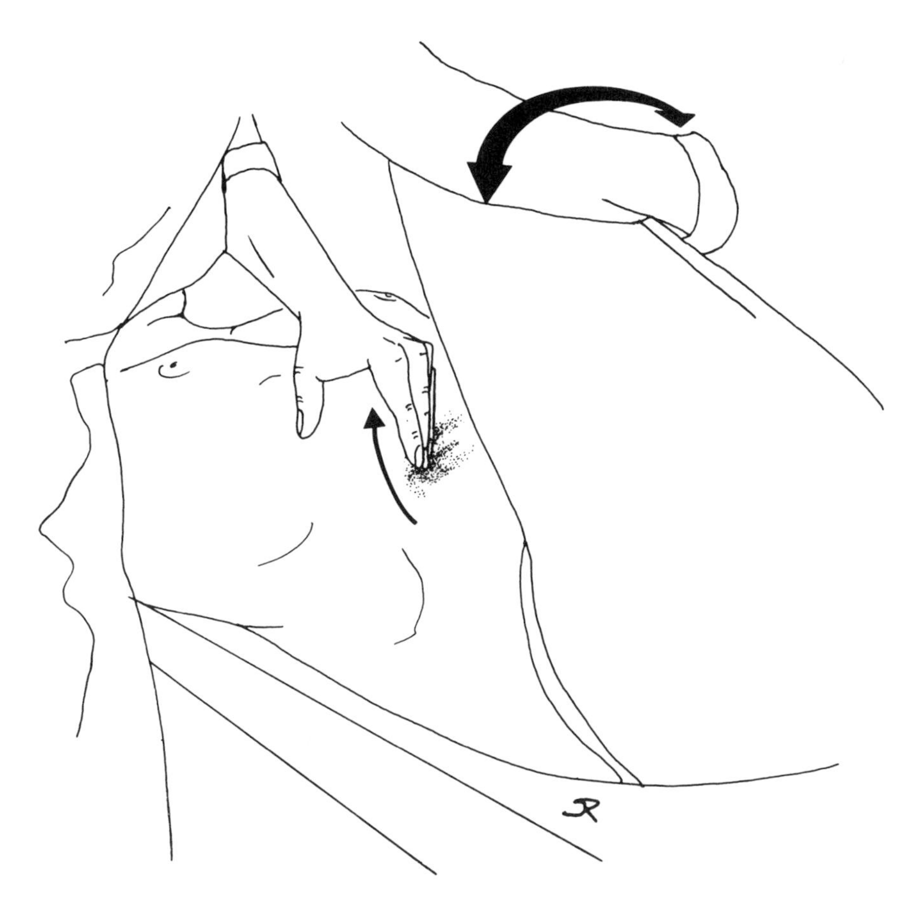

Illustration 4-18
Combined Manipulation of the Liver — Seated Position

The choice between the seated and supine positions will depend on personal preference, circumstances and whatever is more comfortable for you and the patient. However, one or the other may give more optimal results in a particular situation, since the structures through which the long lever is working differ. In the seated position, the long lever is working through the lungs, pericardium and diaphragm; if there are associated problems or restrictions with those tissues, this approach will give the best results. In the supine position, the primary viscera being used as levers are the kidneys and colon; this technique is preferred if there are associated problems or restrictions in these organs.

INDUCTION TECHNIQUES

Liver

These induction techniques are performed in the supine position to give the treating hand the largest possible contact with the right thoracoabdominal surface. Your fingertips should be placed as close to the left triangular ligament as possible, with the hand on the rib cage and its ulnar aspect just extending onto the abdomen.

For learning purposes, it is advisable to go over the motions of motility in the different planes one at a time. We will describe the technique for inducing expir in the different planes. Remember that in induction you work by encouraging the part of the motility cycle which is easiest and has the greatest amplitude. If the motion of inspir were greater, you would then encourage it by doing movements opposite to those described below.

In the frontal plane, the motion of the liver is along the line of the falciform ligament. The superolateral part of the liver moves inferomedially toward the umbilicus during expir *(Illustration 4-19)*. For induction, follow the lateral edge of the liver on the arc

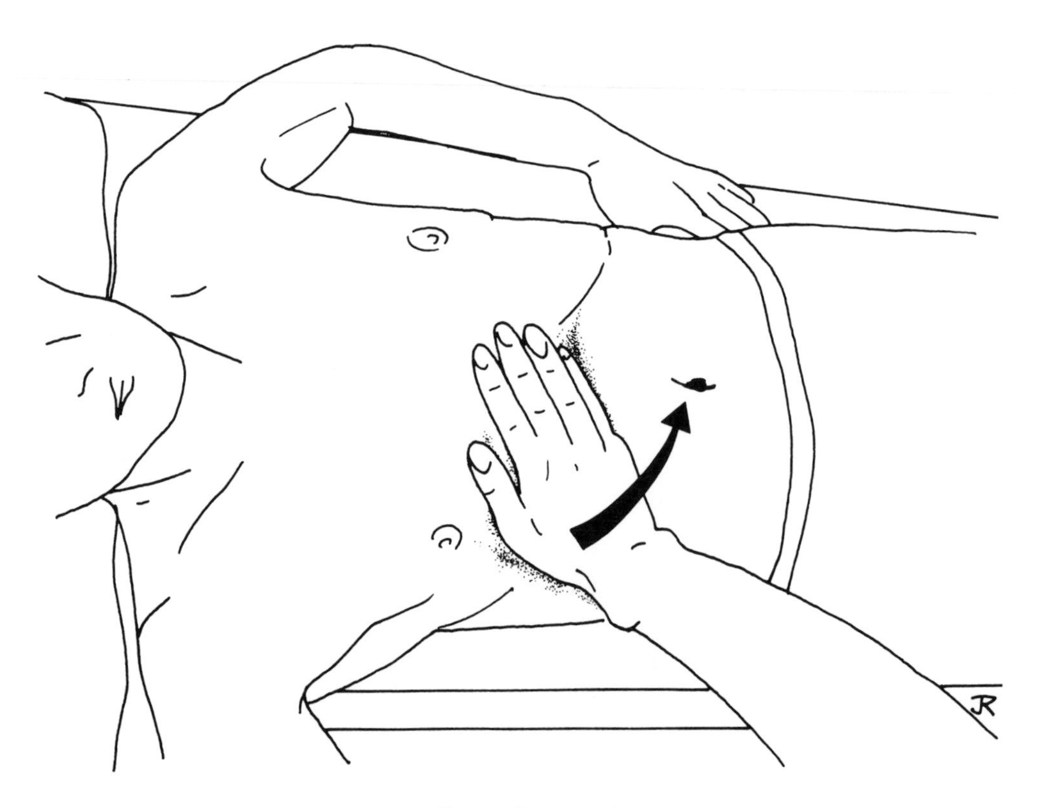

Illustration 4-19
Induction of the Liver in Expir — Frontal Plane

of a circle toward the umbilicus, approximately once every 10 seconds. This is usually the most easily palpable aspect of liver motility and you should familiarize yourself with it. It is important in induction to encourage the motion in precisely the right direction. During motility, if you try to push the organ in the wrong direction, it will stop moving. This stoppage differs from the still point in that as soon as you let go the motion will resume, but with no improvement. The motion of expir is more forgiving than that of inspir.

In a sagittal plane, as the liver turns on a transverse axis during expir, the superior part rolls anteroinferiorly. From the position described above, encourage this motion by supinating the hand on the abdomen *(Illustration 4-20)*. To get a better feel for this motion, try putting one hand posteriorly and one anteriorly, pronating the former and supinating the latter.

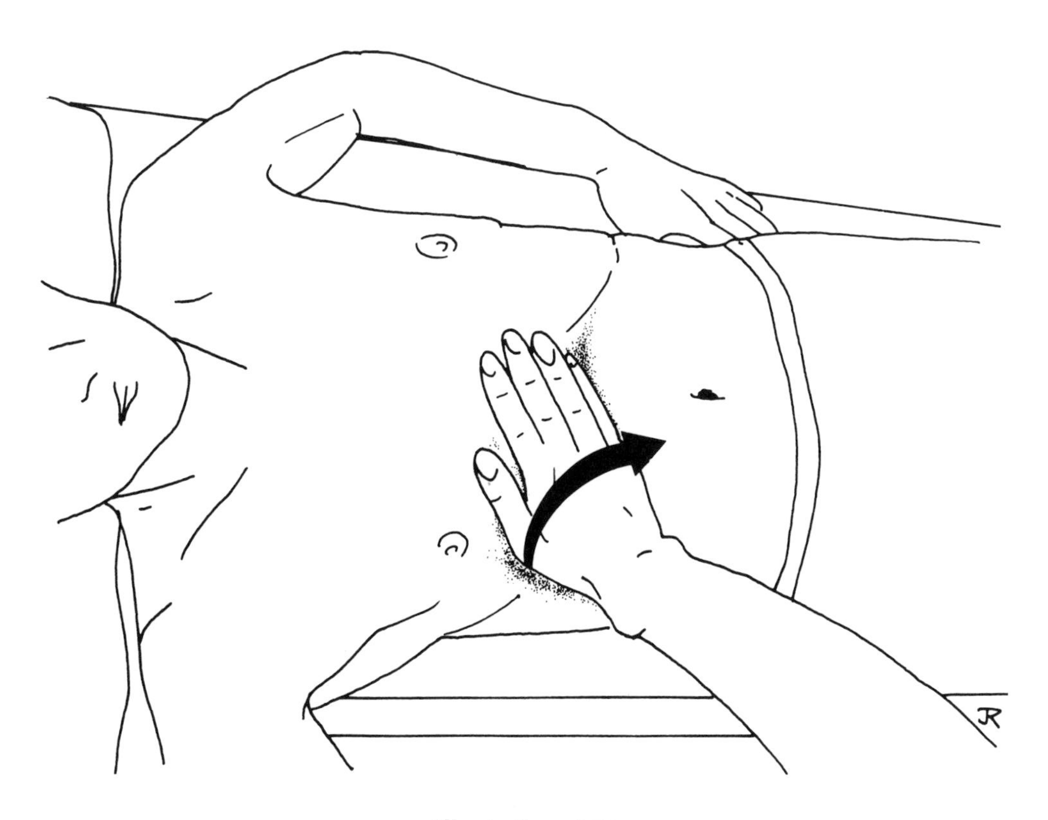

Illustration 4-20
Induction of the Liver in Expir — Sagittal Plane

In a transverse plane, you must follow and encourage the liver's motion in expir around a sagittal axis. From the same position, push or pull the lateral aspect of the rib cage anteromedially with the palm of the hand while your fingers push the medial aspect posteriorly *(Illustration 4-21)*. This accentuates the rotation to the left of the external edge of the liver.

When these induction movements are to be combined, first induce the motion in the frontal plane, as it is the primary motion of this organ and easiest to appreciate.

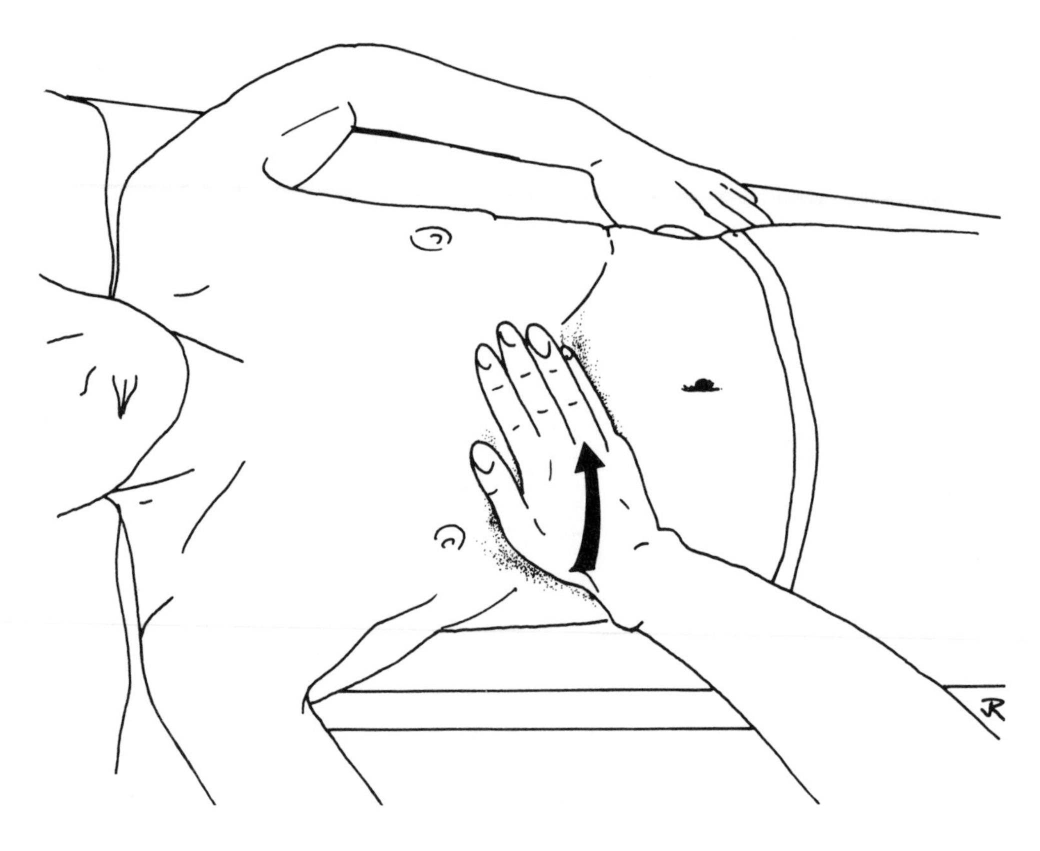

Illustration 4-21
Induction of the Liver in Expir — Transverse Plane

After feeling this confidently, add the motions in the sagittal plane and finally the transverse plane. In clinical practice, this is the most common way to perform induction of the liver.

Common Bile Duct

Induction of the common bile duct is performed in the supine position and is similar to the direct technique described above (page 101). Gently press the heel of your hand or fingers just underneath the rib cage, on the midclavicular-umbilical line. Do not push hard enough to reach the duct; induction does not work if too much force is used. Focus your attention on the anterior projection of the duct. Move your hand very slightly posteromedially and then progressively anteriorly, in a sine wave pattern. Return your hand following the reverse route. At first you push; as the technique progresses you just follow the motion. We advise beginners to put a stethoscope under the hand to listen for evidence of biliary evacuation. This is a good way not only to evaluate a student's performance, but also to convince skeptics of the power of induction.

Effects

The most obvious effects are those of biliary "flushing." The manipulations work on the mechanisms of biliary transit and cause different motions of the various ducts and channels, as we have often seen using fluoroscopy. In our own experiments, correctly performed, these techniques can increase biliary transit by 30%. When performed incorrectly (wrong direction or inappropriate force), they stop it altogether. We believe, based on the reactions of many of our patients, that there are also effects upon the metabolism of the liver, but have not yet been able to demonstrate these objectively.

Following manipulation of the gall bladder, the patient will sometimes experience vomiting and/or fatigue. The vomiting occurs in very short episodes, Often the color of the vomitus will start out being green and turn to yellow. These are transitory effects and last from 1-3 days. When they have passed, the patient will report that he feels great.

When manipulation of the liver is completed, it is important to continue down the bile duct all the way to Oddi's sphincter. The manipulation of this sphincter is discussed in chapter 5. Successful manipulation of the liver requires that the adjacent organs (stomach, colon, duodenum and kidney) also have healthy mobility and motility. All the organs must articulate harmoniously for any one of them to be said to move well.

Adjunctive Considerations

ASSOCIATED OSSEOUS RESTRICTIONS

There are often restrictions in the vertebral zones corresponding to the liver — T7-9 and C4/5. The transverse process of C4, particularly on the left, is a trigger point to release the gall bladder. This is primarily due to a link via the phrenic nerve. There is a right scapulary hepatic projection due to another branch of the phrenic nerve. Because of a direct relationship, as well as the mechanical tensions of the liver attachments, the lower ribs and right costovertebral articulations can also be affected. These are classical zones and the fact that they have been described does not mean they need to be manipulated — that would be forgetting the major concept of the total lesion. If restriction of the liver is the causal problem, proper manipulation of it will lead to a disappearance of the secondary vertebral restrictions.

RECOMMENDATIONS

Patients appreciate the reassurance of hearing again what they already know. Besides warning them to avoid fatty substances, sugar and alcoholic drinks, you should advise them to increase their intake of warm or hot fluids between meals. Cold drinks are forbidden as cold slows down the process of digestion by inhibiting the effects of enzymes. Our patients report that cold liquids have an especially deleterious effect in the afternoon. Effective stimulants of the hepatic and biliary systems, following osteopathic treatment, include: (1) a 10-day course of lemon juice in lukewarm water with a papaya or a teaspoon of olive oil in the morning before breakfast; (2) artichokes; and (3) bitter salads. Hot packs over the liver are a good adjunctive technique.

We have noticed that women often experience an increased sensitivity to certain foods (such as chocolate and white wine) during the second half of the menstrual cycle.

We believe that certain reproductive hormones increase the effects of tyramine on the brain. This could explain why at certain times women can eat anything, while at other times the slightest trace of tyramine in a food causes illness.

Try to find out which particular foods affect your patient. One method we have used is to place the food in question on the patient's abdomen, in small bottles, and then test the motility of the liver. In some cases, the motility will diminish or even stop. There is no clear explanation for this phenomenon. Is it a psychological or a physical effect? When we use this method, the patient does not know what we have placed on her abdomen. When we use empty (control) bottles there is consistently no change in the motility.

It is interesting to note that people are often attracted to the substances which affect them adversely. Perhaps the absorption of these substances leads to euphoria via the secretion of endorphins — sadly the aftermath is not so pleasant!

Chapter Five: The Esophagus and Stomach

Table of Contents

CHAPTER FIVE

The Esophagus and Stomach

The mechanics of the gastroesophageal junction are very complex. This is an area where many conflicting forces, including mechanical traction, meet. The esophagus and upper part of the stomach are torn between the negative pressure of the thorax and the positive pressure of the abdomen, leading to a variety of mechanical pathologies in this region. The upper part of the stomach is drawn upward, causing risk of hiatus hernia, while the middle and lower parts are drawn downward, causing risk of gastric prolapse. Because this permanent stress exists even under normal conditions, a variety of pathologies and associated symptoms can result.

Anatomy

The thoracic portion of the esophagus is found in the posterior mediastinum and is closely linked to the trachea by connective tissue. The esophagus deviates to the left, and is linked to the left bronchus. It often adheres to the pleura, and comes in contact with the pericardium. Above the level of T4, it presses on the vertebral column, after which it moves away from the spine and is separated from it by the aorta at the level of T7 or T8.

The diaphragmatic portion of the esophagus is 2cm long, the anterior side being covered by the peritoneum and making an indentation on the posterior side of the liver. The posterior side leans directly on the left crus of the diaphragm.

The stomach is situated in the supramesocolic cavity and occupies the largest part of the subphrenic space. The distance from the gastric fundus to the pyloric antrum is normally 25cm, and that from the lesser to the greater curvature 12cm. The stomach's average capacity is 1,200ml. Its form varies greatly from individual to individual and according to biotype — for example, in an asthenic subject, the pyloric antrum can be anywhere from the level of L3 to the pubic symphysis.

RELATIONSHIPS

The thoracic esophagus contacts the trachea, left bronchus, pleura and pericardium. When there are mechanical problems affecting the esophagus (e.g., reflux or hiatus hernia), there is often pain in the cardiac area induced by abnormal tension of the tissues connecting the esophagus and pericardium.

Posteriorly, the esophagus (along with the vertebral column and the prevertebral aponeurosis and muscles) is accompanied by the two vagus nerves. The diaphragmatic esophagus contacts the left diaphragmatic crus, aorta and, posteriorly and on the right, the lower part of the left lung, T10 and T11. On the left, the left triangular ligament of the liver is continuous with the parietal peritoneum covering the diaphragm and the esophageal peritoneum; on the right its edge is followed by the lesser omentum.

The orientation of the stomach changes depending on how full it is. The surfaces that are anterior and posterior in the full stomach become more superior and inferior respectively when the stomach is empty.

The left part of the anterosuperior surface relates to the diaphragm and thereby to the pleura, left lung, R6-9 and associated costal cartilages. The right part of this surface is mostly covered by the left and quadrate lobes of the liver, or is directly in contact with the anterior abdominal wall, where it is easily percussed *(Illustration 5-1)*.

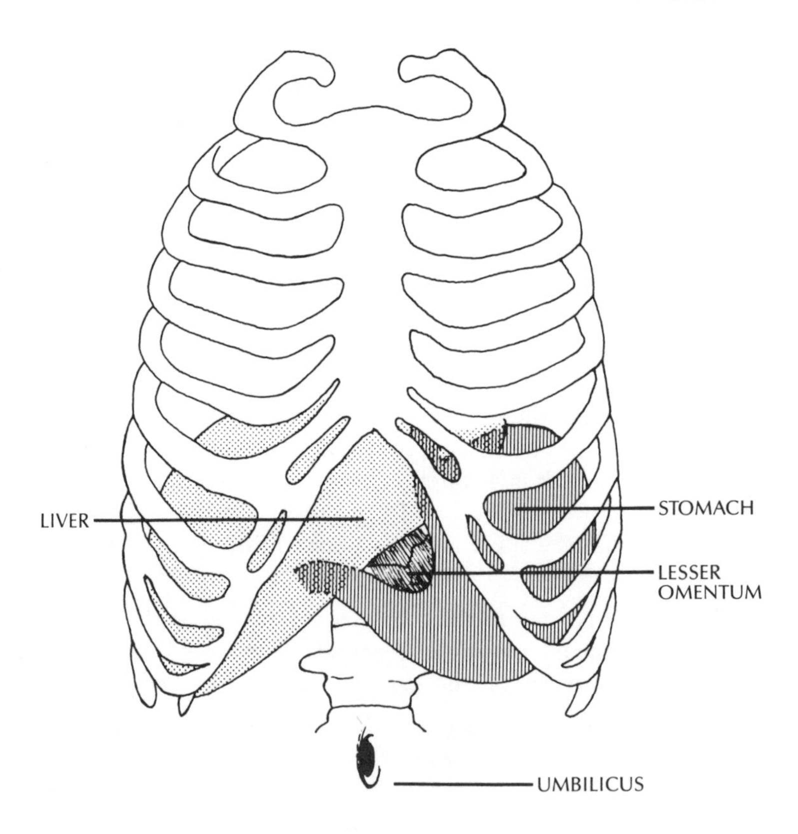

Illustration 5-1
Anatomical Relationships of the Stomach

The posterosuperior surface of the stomach is linked with the diaphragm by the gastrophrenic ligament. Except for a small area near the cardiac orifice which directly contacts the left crus of the diaphragm, this surface is covered with peritoneum. It relates to the left adrenal gland, anterior part of the pancreas, left kidney, left colic flexure and upper layer of the transverse mesocolon. This mesocolon and the greater omentum separate the stomach from the small intestine and duodenojejunal flexure.

The lesser curvature is attached to the vertebral column (from T1 to L1), celiac trunk, caudate lobe of the liver and solar plexus. The greater curvature is connected to the diaphragm by the gastrophrenic ligament, spleen, transverse colon and greater omentum.

The pylorus is related to the median or left part of the body of L3; it also relates anteriorly with the inferior side of the liver, posteriorly with the portal vein and hepatic artery, superiorly with the lesser omentum and inferiorly with the head of the pancreas.

VISCERAL ARTICULATIONS

The many articulations discussed here underline the existence of complex interrelationships in the human body, and the fact that visceral restrictions can have repercussions which are sometimes not easy to understand. As with the other viscera, turgor and intracavitary pressure are essential for the support and cohesion of the stomach.

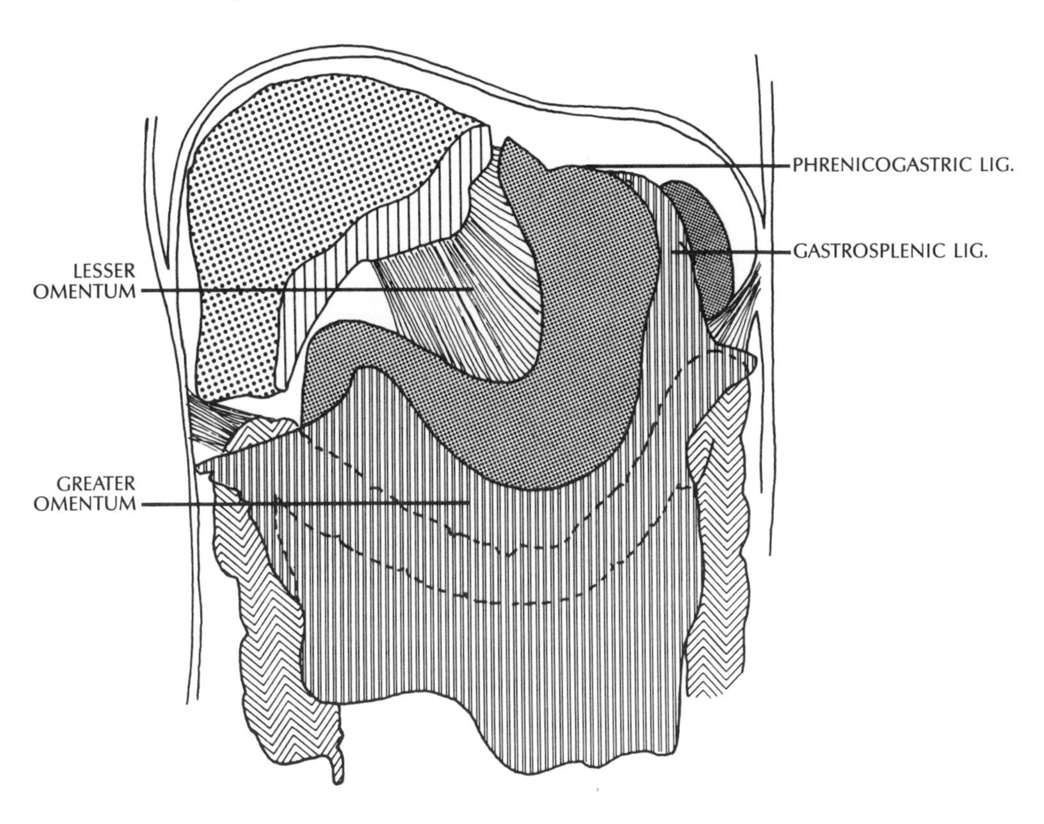

Illustration 5-2
Visceral Articulations of the Stomach

The esophagus is joined to the trachea by connective tissue and is sometimes attached to the pleura. When crossing the diaphragm, it is covered by a fibrous subperitoneal sheath which is attached to the peritoneum and to the diaphragmatic crura *(Delmas 1975)*. There are muscular fibers between the diaphragm and esophagus which reinforce this sheath.

The powerful gastrophrenic ligament, which is really the supporting ligament of the stomach, connects the gastric fundus to the greater curvature to the diaphragm. It is a continuation of the coronary ligament which supports the liver.

The lesser omentum joins the lesser curvature of the stomach to the liver. It is found far posteriorly and faces right and superiorly. The greater omentum is a peritoneal fold which joins the stomach to the transverse colon. It is joined to the diaphragm by the phrenicocolic ligaments at the level of the colonic flexures. The gastrosplenic omentum connects the stomach to the spleen but does not appear to have a supporting role *(Illustration 5-2)*.

To summarize, the stomach is interdependent with the diaphragm by means of the gastrophrenic ligament and greater omentum; it also has a very close connection with the liver via the lesser omentum.

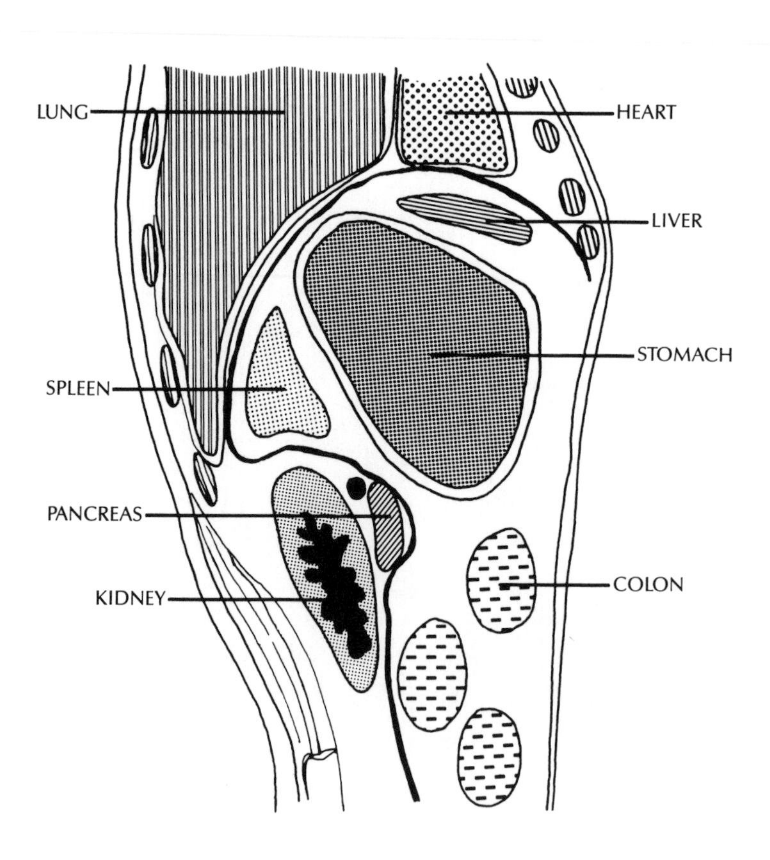

Illustration 5-3
Sliding Surfaces of the Stomach

Sliding Surfaces

The stomach articulates with the diaphragm and thereby with the heart, pericardium, lung and left pleura. One can really speak of articulations here, as harmony of movement between these organs is very important. It also articulates with the liver, where it leaves a large impression on the left lobe. Remember that the left lobe and the left triangular ligament penetrate between the diaphragm and the anterior side of the stomach. The left upper part of the liver is in front of the stomach. The stomach also articulates, directly or indirectly, with the splenic angle of the colon, spleen, pancreas, transverse colon and its mesocolon, inferior duodenum and the left kidney and adrenal gland *(Illustration 5-3).* In some cases, when the stomach is atonic and prolapsed, it slides on the colon, small intestine and sometimes even the bladder.

TOPOGRAPHICAL ANATOMY

The anterior part of the stomach stretches from the left edge of the sternum to the adjacent left part of the thorax, and at the top from the 5th intercostal space to the left inferior edge of the thorax. The cardia is located 2cm from the median line at the level of T11 posteriorly and the 7th costal cartilage anteriorly. The lesser curvature goes

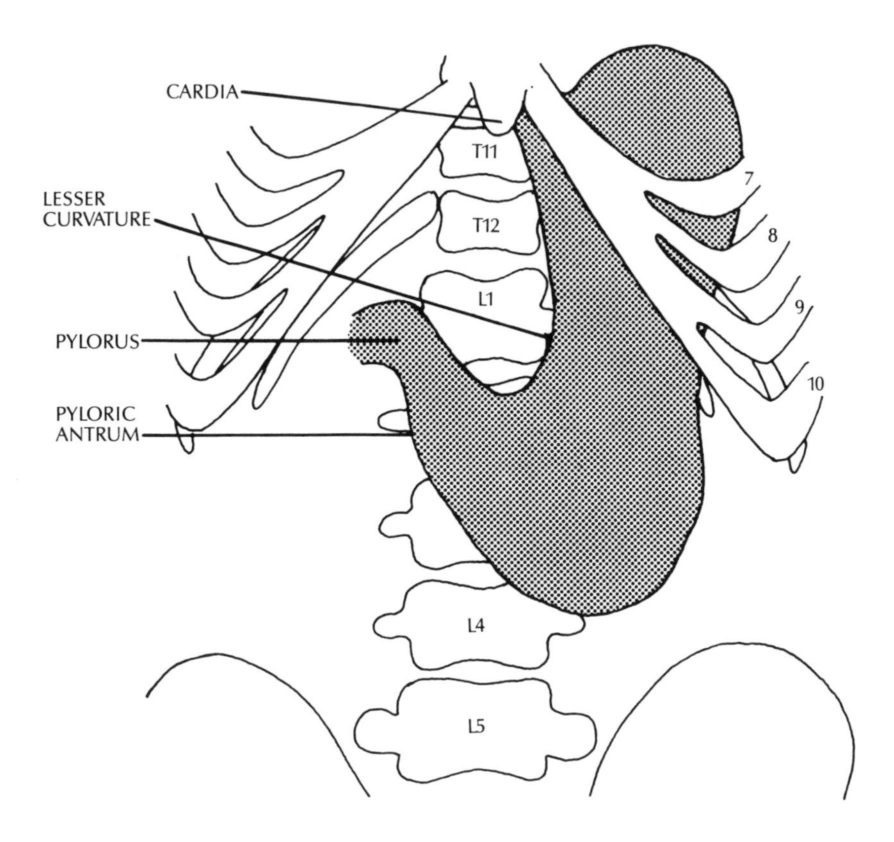

Illustration 5-4
Topographical Anatomy of the Stomach

from this cartilage to the left lateral aspect of L1; from T10 to T11 it relates to the vertebral column. The greater curvature follows the interior aspect of the anterior face of the ribs.

The pyloric antrum is found on the left lateral aspect of L2/3. In theory, it is always below the umbilicus; in fact, its position varies widely. The pylorus, like the cardia, is deeply situated. When the stomach is empty, it is found slightly to the left of the median line, 6-7cm above the umbilicus; when full, it moves 1-2cm downward and 3-4cm to the right. When the subject is standing, the pylorus moves to the median or left part of the body of L3; when lying down, it is related to L1-L2 *(Illustration 5-4).*

Physiologic Motion

ESOPHAGUS

There is a longitudinal tension of the esophagus which stabilizes it and encourages alimentary transit. This tension contributes to the closing of the inferior part of the esophagus because, at the level of the cardia, the esophagus rotates on its axis and thereby forms an elastic, twisting occlusion which contributes to the functional sphincter of the cardia. This occlusive effect is reinforced not only by abdominal pressure, but also by the presence of a mattress of submucosal veins. During inhalation, the lower part of the thoracic esophagus moves to a point as much as 7cm away from the vertebral column; at the level of the esophageal orifice, the esophagus remains longitudinally mobile inside its fibromuscular sheath. This sheath is attached to the diaphragmatic crura and peritoneum and there is a cellular sliding space in between. It is, therefore, effectively a true muscular canal. The thoracic part of the esophagus undergoes pulmonary traction, whereas the abdominal part is influenced by abdominal pressure. The upper part always wins; the thorax, like a magnet, attracts all the viscera that adhere to the diaphragm.

STOMACH

Mobility

While crossing the diaphragm, the esophagus slides in a fibromuscular sheath. In contrast, the stomach, being closely attached to the diaphragm by the gastric fundus, moves with it. We shall describe, on the three different planes, the movements of the stomach during inhalation.

On a frontal plane, the phrenic center is lowered, but much less so than the left part of the diaphragm. Because of the structure and disposition of the diaphragm, its posterolateral aspect moves most (remember its sagittal concavity). Therefore, the gastric fundus is moved inferomedially. The distance between the lesser and greater curvatures decreases, as does the distance between the gastric fundus and pyloric antrum, because the latter moves superiorly and to the right. At the end of the movement, the stomach reduces its width toward the medial edge, but because the body of the stomach moves far down during inhalation, the large vertical median axis lengthens. This movement of the body is more important than that of the gastric fundus. The stomach side-bends to the left (moves clockwise when viewed from the front), and it is always the gastric fundus, greater curvature and body which move the most. The anteroposterior axis passes through the lower part of the lesser curvature (the angular notch) near the inferior insertion of the falciform ligament *(Illustration 5-5).*

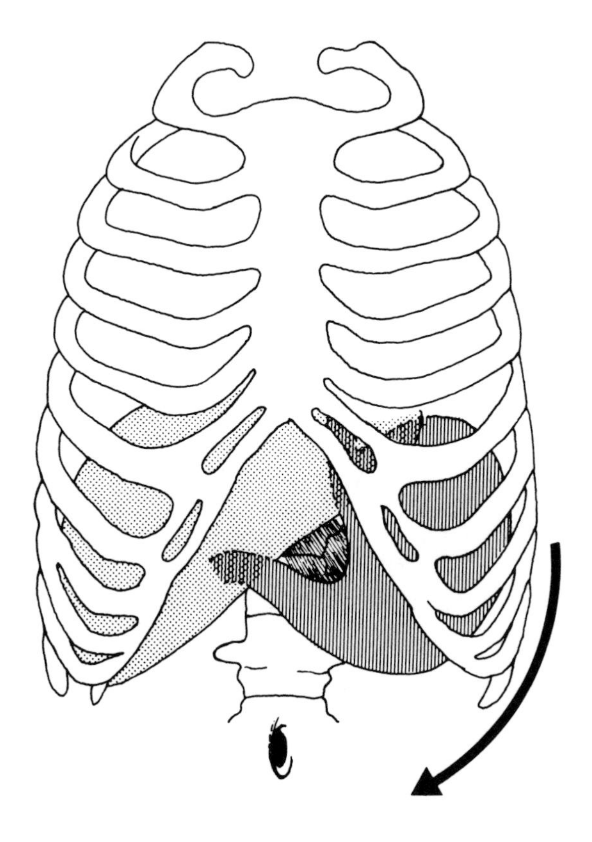

Illustration 5-5
Mobility and Motility of the Stomach — Frontal Plane

In a sagittal plane, the stomach can be easily deformed and its movement is more difficult to analyze than that of the liver. A swing from back to front is accomplished in such a manner that the gastric fundus dives forward while the pyloric antrum moves backward. The axis goes through the middle of the stomach. It is mostly the upper part that moves *(Illustration 5-6)*.

In a transverse plane, when the diaphragm is lowered, the gastric fundus is forced (because of the anchoring points of the esophagus) into a right rotation following a vertical axis that goes through the inferior part of the esophagus *(Illustration 5-7)*.

Motility

The movements of motility are similar to those discussed above, but with a different rhythm and amplitude. We shall briefly discuss what happens during expir.

In a frontal plane, the gastric fundus and greater curvature lower themselves and sidebend to the left. This sidebending is one of the most important movements in visceral listening *(Illustration 5-5)*.

In a sagittal plane, the forward diving movement of the gastric fundus is only slightly palpable; it is very difficult for beginners to perceive *(Illustration 5-6)*.

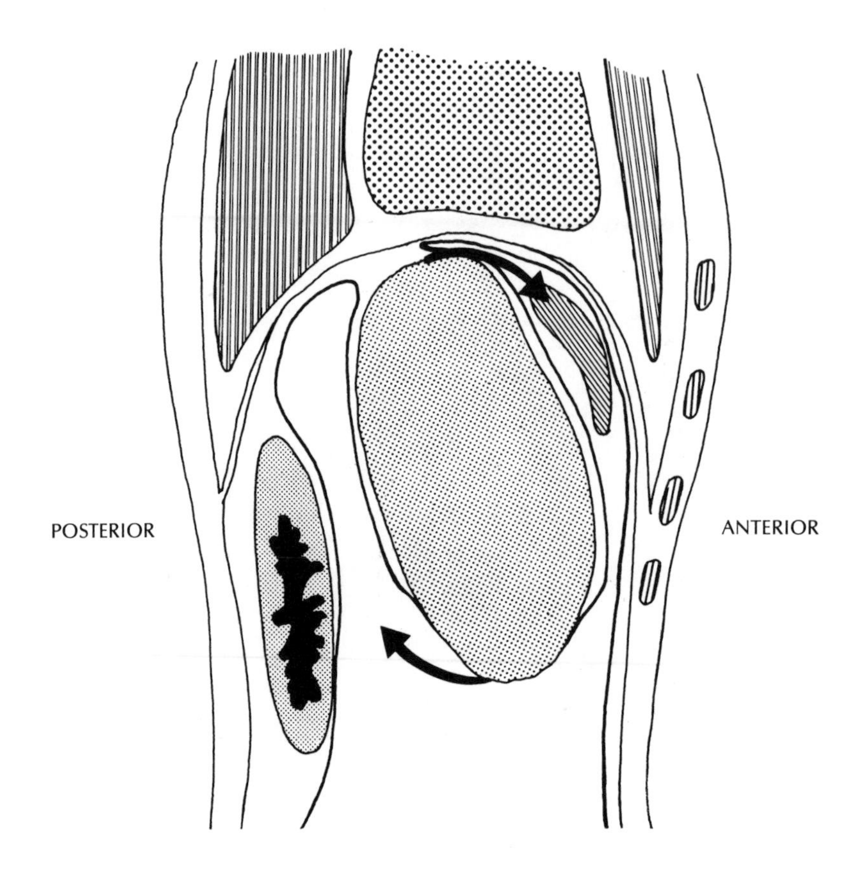

Illustration 5-6
Mobility and Motility of the Stomach — Sagittal Plane

In a transverse plane, the right rotation of the stomach is important for listening movements; the anterior, medial and inferior motion of the greater curvature is easily felt. During the process of induction, achieving a good amplitude and quality of this movement is essential for obtaining good motility overall *(Illustration 5-7).*

Indications for Visceral Evaluation

As stated in the preceding chapters, it is not feasible to cover all possible indications for evaluation of a given area. Osteopathy is one of those medical fields in which the practitioner remains a whole — she is not a technician that can be replaced by a machine. No one else can do what she does, in exactly the same way. Take up these indications and techniques and integrate them into your therapeutic kit — as soon as you use them, they will become yours.

We have divided the syndromes for which an evaluation of gastroesophageal motion is particularly warranted into two groups: mechanical and irritative. Obviously, the barrier between the two is for didactic purposes only.

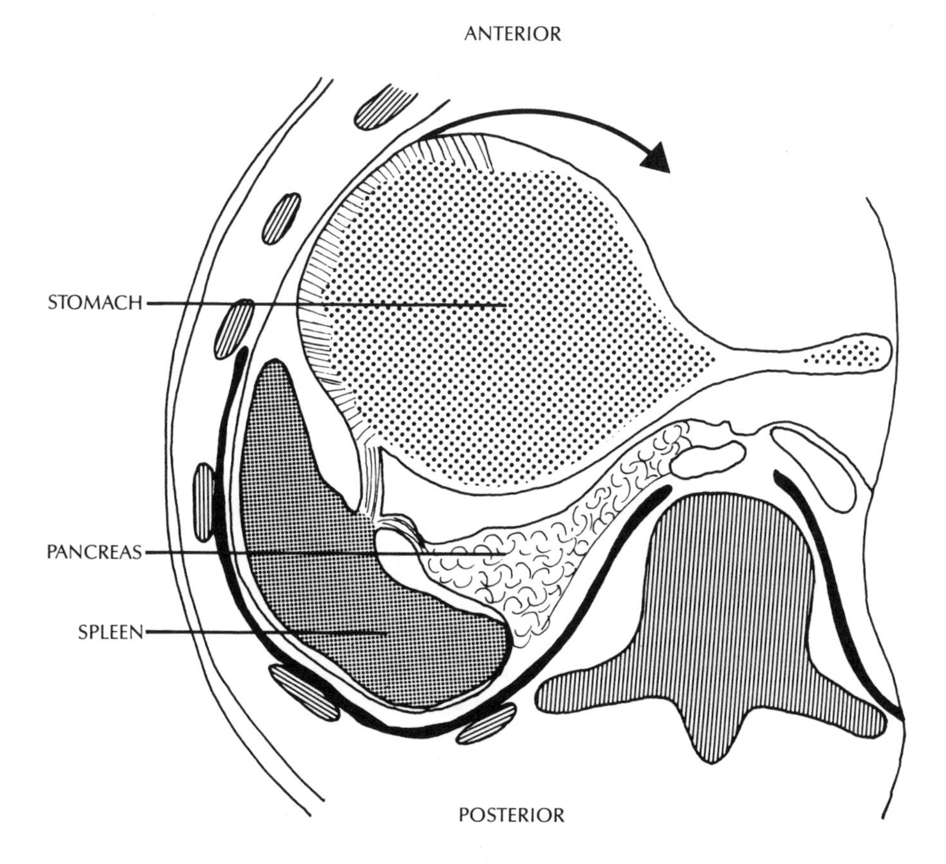

Illustration 5-7
Mobility and Motility of the Stomach — Transverse Plane

MECHANICAL SYNDROMES

Gastroesophageal Junction

A problem here is really a good indication for visceral manipulation. We have already seen that the functional sphincter of the cardia relies upon a balanced tension between the suprahiatal and the subhiatal forces of traction. If this balance is broken, the axial rotation of the esophagus, which contributes to cardial occlusion, no longer works well and gastric reflux is likely to occur.

The suprahiatal pressures and the traction exerted by the thorax on the diaphragm literally suck the stomach upward — hence the important risk of hiatus hernias. The principal causes of hiatus hernias are mechanical, e.g., kyphosis, hypotonia, gastric ptosis and obesity. Contributing factors include general debility and aging, which weaken the tissues and decrease tone. All pleural problems can lead to problems of the hiatal area. Because an excess of estrogen or decrease in progesterone can loosen up the hiatus, these problems are common during pregnancy and at menopause. These factors all contribute to the displacement of the hiatal center. The slightest displacement of this center means that superior negative pressures will break the balance of the cardia at the expense of the stomach — we have never seen an esophagus which forms a hernia in the abdominal cavity! This confirms again that in the upper part of the abdomen, pressure is from bottom to top. Hiatus hernia is quite common; it occurs in over 60% of people over 60 years old. Our methods of manipulation are very effective in these cases.

With certain pharyngeal or laryngeal irritations, one must consider the phenomenon of gastric reflux. In such cases, the patient complains of atypical sore throats with no visible cause. All the ENT examinations have negative results and the patient may be mistakenly diagnosed as being depressed. Usually manipulations of the hiatus solve this functional problem rapidly. Even if in radiography the hernia persists, the cardia rediscovers its function and there is no significant reflux.

Gastric Ptosis

Radiography will often show the pyloric antrum lower down than it is usually described in anatomy books. With longitudinal asthenics, it is not unusual to find it at the level of the pubic symphysis. Most often it is only the pyloric antrum which collapses; the gastric fundus remains in place. In a real ptosis the gastric fundus is also lowered, but because the word is of common usage, we will use the term "gastric ptosis" for even a simple lengthening of the stomach.

The stomach is subjected to conflicting mechanical tensions — the superior part is strongly drawn upward while the inferior part, particularly if placed far down, is drawn toward the feet. If the stomach is in its normal supraumbilical position the effect of this conflict is negligible, but if the pyloric antrum falls lower than the umbilicus, the effect becomes more significant. At the level of the fundus, there are cholinergic fibers which, if they are stretched too much, increase the secretion of hydrochloric acid. Apart from congenital causes, most causes of gastric ptosis relate to: hypotonia; kyphosis (which brings the diaphragm closer to the pubis); hard work with arms raised; multiparity; age (which causes the gastric fibers to decrease in tone); and certain hormonal disorders.

A note about the problem of people who work with their arms raised (e.g., painters): It is often observed that painters suffer from gastralgia, mostly due to inhalation of toxic products, but we were surprised to ascertain that gastralgia increases when painting on the ceiling! The "arms up" position noticeably increases the vertical length of the stomach; the cholinergic fundal fibers are stretched and secrete hydrochloric acid. Conceivably, the vagus nerves are also affected, by the stretching of the thorax and the posterior flexed position of the cervical vertebrae.

IRRITATIVE SYNDROMES

These include all attacks on the gastric mucosa, from a simple gastritis to an ulcer, for either psychological, dietary, chemical or infectious reasons. The aim of the manipulations is to stop the visceral spasm of the gastric fibers or the pyloric sphincter (the duodenum will be discussed in chapter 6), and to decrease the gastric transit time to avoid stagnation of either acid or relatively alkaline liquids. The stomach must mix its contents so that there is maximal contact between the food to be digested and the gastric mucosa. When the stomach is atonic, certain parts of the lining stay in contact with sugars and other food particles too long, which causes irritation. On the other hand, other parts are never in contact with the food and therefore never carry out their role in the digestive process which, in turn, slows down digestion even more. Manipulations of the pylorus and duodenum are important because dysfunction of the pyloric sphincter will interfere with the circulation of liquids (particularly bile), which can reflux into the stomach and causes problems, particularly when stagnant. As transit time in the stomach increases, its pocket of air expands, provoking the wellknown associated cardiorespiratory symptoms.

Evaluation

The history is focused on abdominal pains, their quality and rhythm. When recurring pain is present, your task is to differentiate between simple gastritis, ulcer and neoplasm. A perforated ulcer is a medical emergency and you must prescribe the necessary tests or studies if there is any doubt.

Gastric ptosis is characterized by flatulence, heaviness, eructation and difficulty in wearing a belt or sleeping on the stomach. The peculiarity of discomfort due to ptosis is that it occurs as the stomach fills: around ten minutes into eating or soon after drinking quickly. People with this condition feel very hungry when they eat, but the hunger fades quickly. As their gastric transit time is increased, you can often hear a sloshing sound in the abdomen. Pushing down on the stomach may recreate the symptoms. There is often a slight anemia due to reduced gastrin secretion.

You should also be familiar with the symptoms of hiatus hernia. These include very deep pain at the level of T11 that increases with breathing, an increase of air in the stomach, heartburn, difficulty in swallowing, left-sided chest pain and pyrosis. Pushing the stomach upward will often recreate the symptoms. Reflux can have many effects on children, most particularly asthma (late at night or very early in the morning) and tonsillitis. In recurrent, seemingly inexplicable cases of this sort, it is worthwhile to check the mobility and motility of the stomach.

There are other distant symptoms which can have their origin in the stomach. Two of the most common are left frontal headaches and left shoulder pain. For a description of a simple way to determine if a stomach dysfunction is contributing to a periarthritis of the right shoulder, see page 91.

Percussion of the stomach is an important part of the exam. It is performed where the anterior face of the stomach is in contact with the anterior abdominal wall, and can help delineate the stomach's position. The top of the stomach should reach R6/7 at the midclavicular line. We shall talk about palpation of the stomach in the following section.

If there is any doubt, do not hesitate to order a radiograph or refer for endoscopy of the stomach. Even the effects of cumulative amounts of radiation are nothing compared to the catastrophic impacts of a perforated ulcer or undetected gastric cancer.

MOBILITY TESTS

For mobility tests, our preferred approach (and the only one which enables us to go deeply into the stomach) is the direct subcostal route with the patient sitting on the examination table, legs hanging down and the back kyphosed to release abdominal tension. It is obvious that abdominal pressure is different when standing up than when lying down. With the seated position, the stomach is in its normal position and pulls on the insertions, accentuating any possible restrictions.

Stand behind the patient, with your fingers placed under the left costal edge and then move them posterosuperiorly as you progressively forward bend the patient. In this way, you can push your fingers as much as 10cm toward the back without causing pain *(Illustration 5-8)*.

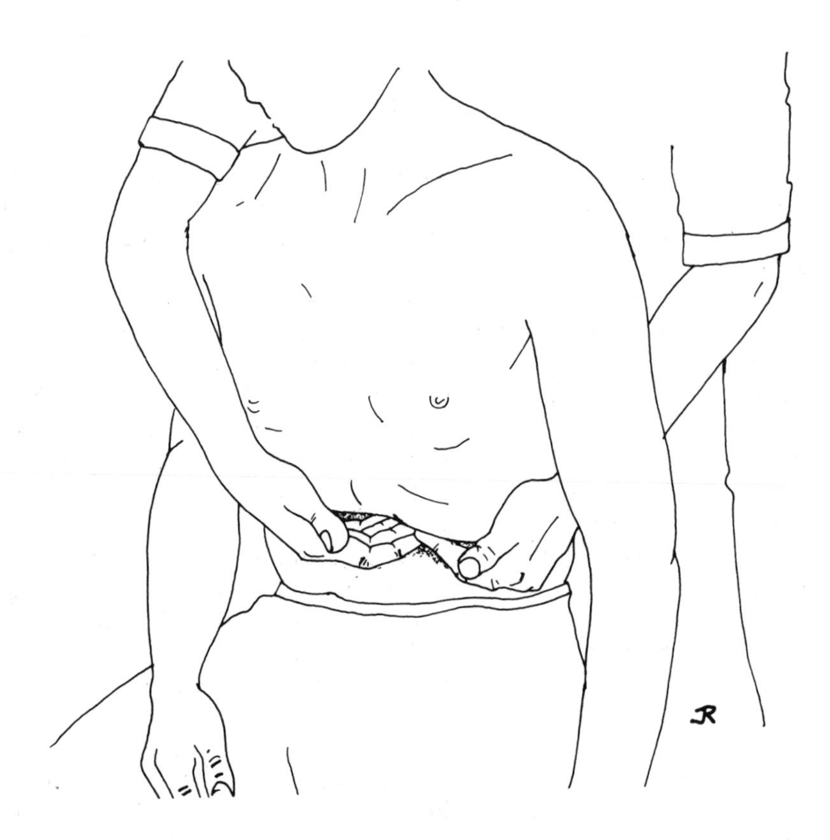

Illustration 5-8
Direct Subcostal Approach to the Stomach — Seated Position

From right to left, the first organ you will feel is the transverse colon, followed immediately by the stomach. If you place your fingers as far as possible to the left and upward, they will contact the splenic flexure and left phrenicocolic ligament. Further to the right and upward, you will feel the left extremity of the liver and its left triangular ligament. It may be difficult at the beginning to interpret what you feel, but experience will alter this. With a thin patient, you should feel the lesser curvature and the lesser omentum medially. If you feel aortic pulsations, move the fingers slightly to the right.

The pylorus is usually felt approximately 6-7cm above the umbilicus when the stomach is empty. With contraction, it moves from left to right across the midline. As the patient stands, it normally drops 2-3cm. Other possible variations in its position are described above (page 116). As a sphincter, it is tighter than the surrounding area, which should be readily apparent on palpation.

The mobility test for the superior gastric fundus is carried out with the patient in a kyphosed seated position. Standing behind the patient, place the ulnar aspects of your fingers directly subcostal, 3-4cm below the xiphoid and one finger's width under the costal edge. Push your fingers posteriorly and, when they can go no further, direct them superiorly. The pressure must not cause pain. The gastric fundus is at the level of R5/6 on the left and the technique consists of pulling the stomach superolaterally. With a restriction, this movement will take some time to carry out. The stomach can be fixed at the bottom (ptosis) or at the back (in its relationships with the kidney, small intestine, spleen, etc.). This test is most effective for quantifying the size of the air pocket. Although they may feel similar, do not mistake air in the stomach for air in the left part of the colon.

Alternatively, place the patient in a right lateral decubitus position and stand behind him with your fingers under the costal edge of the left midclavicular-umbilical line. Use the same technique as above *(Illustration 5-9)*. Although it is difficult to probe as deeply as in the seated position, the stomach is directed laterally and then superiorly when in contact with the fingers. This method can be easier on both you and the patient.

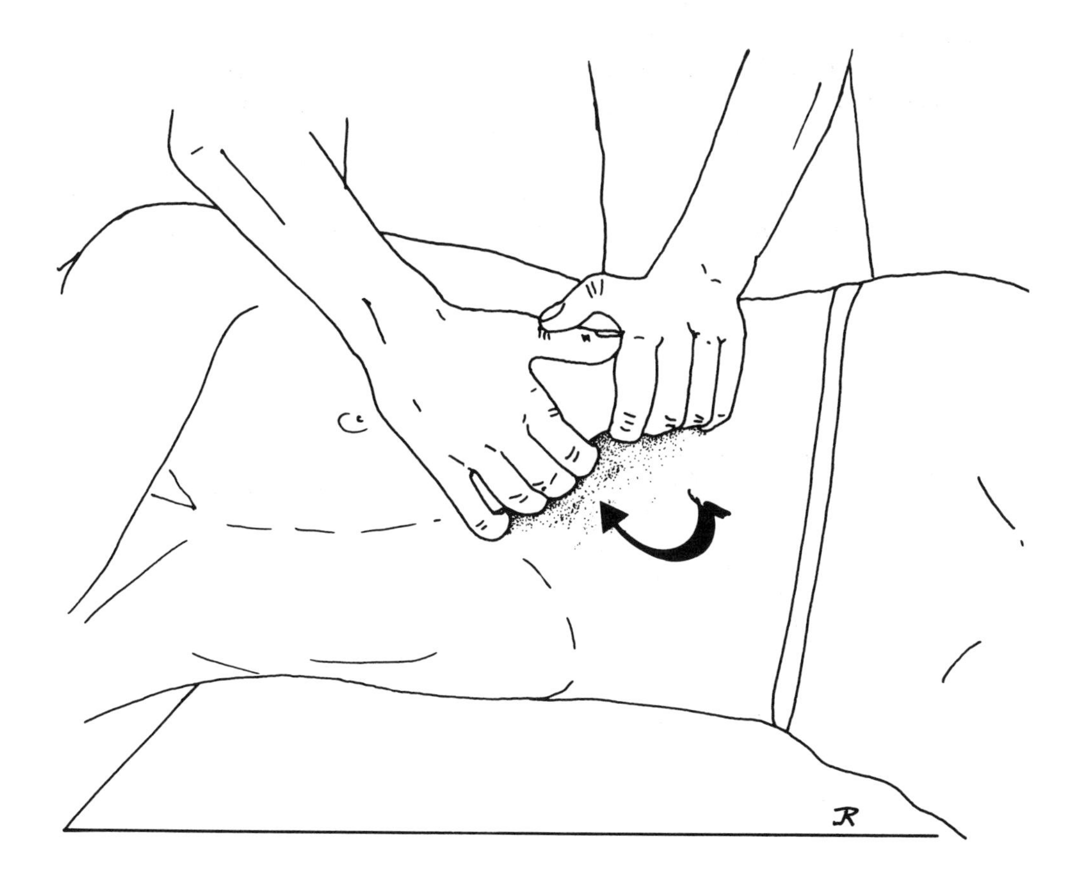

Illustration 5-9
Mobility Test of the Gastric Fundus in the Right Lateral Decubitus Position

The mobility test for the lesser curvature and pylorus, with the patient seated, involves passing by the anterior part of the stomach or the liver. This deep area is more difficult to reach and mobilize. The position of the pylorus may vary from right to left depending on how full the stomach is. Place your fingers under the 7th right or left costochondral junction (depending where the pylorus is) and direct them obliquely upward and to the right — you will draw out the liver or the anterior portion of the stomach and then the lesser curvature and pylorus. The results of this test are difficult to appreciate and you should not attempt it until you are experienced in mobility testing.

MOTILITY TESTS

Place the patient in a supine position, your right hand flat on his abdomen so that the little finger and ulnar edge of your hand are on the greater curvature, the pisiform in the area of the pyloric antrum, the thumb in contact with the duodenum and the index finger following the lesser curvature. The hand is on an axis which goes obliquely downward and to the right. This is the basic position for all manipulations of stomach motility. Obviously, the hand is not only in contact with the stomach; underneath it are the lower ribs, transverse colon, greater omentum, etc.

In a frontal plane, during expir, you should feel your hand moving clockwise around an anteroposterior axis passing near the knuckle of the index finger. The fingers move slightly to the left and inferiorly, whereas the thumb moves superiorly and to the right. Because of this rotation, the stomach shortens in a vertical direction and the palm moves closer to the xiphoid *(Illustration 5-10)*. The opposite happens during inspir. If you find this test difficult to perform, look at your hand and visualize its relation to the stomach.

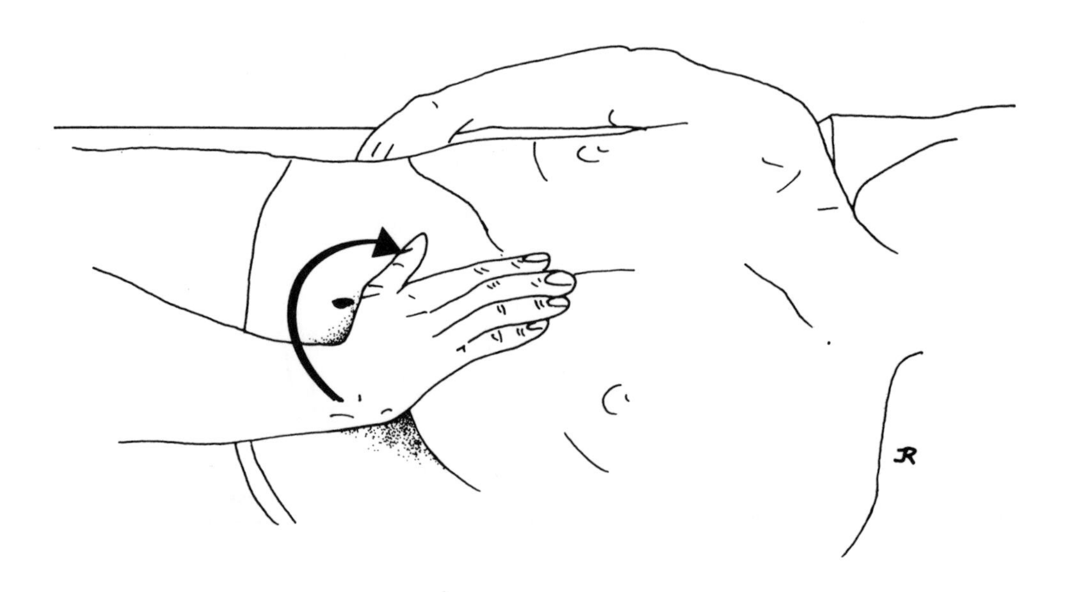

Illustration 5-10
Motility Test of the Stomach — Frontal Plane

To check motion in a sagittal plane, your fingertips should move anteriorly while your palm penetrates posteriorly into the body.

To check motion in a transverse plane, your fingers should move anterolaterally while your thumb penetrates slightly. The vertical axis passes deep to the index finger.

Once you have felt each of the above motions, plane by plane, you should integrate them into one smooth motion *(Illustration 5-11).* They have been divided above strictly for didactic purposes.

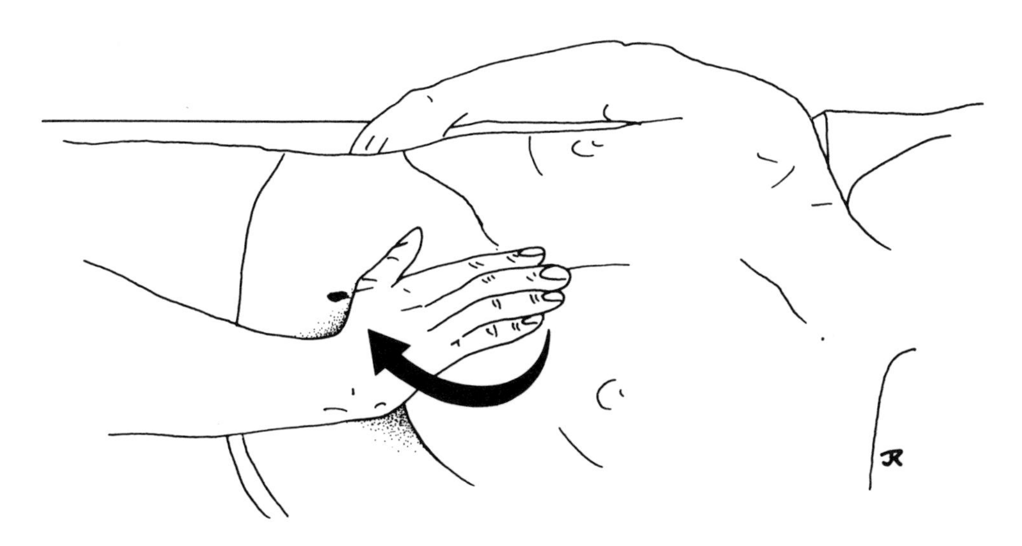

Illustration 5-11
Motility Tests of the Stomach — Resultant Motion

Manipulations

DIRECT TECHNIQUES

Gastroesophageal Junction

This technique is carried out with the patient in a kyphosed seated position which permits deep penetration into the subcostal region. It is for those suffering from a hiatus hernia, reflux or pyrosis; the gastroesophageal junction is not functioning either because the diaphragm is pulling abnormally on the hiatus or the stomach is being drawn upward by intrathoracic pressure. These problems are frequent. Diaphragmatic spasms play an important part in these mechanical problems (page 122).

These lesions are always due to the inferior aspects being pulled up and the technique consists of pushing the cardia back downward. You should use a direct subcostal approach, pushing your fingers toward the back 2cm to the left of the linea alba, which will fix the stomach at the back. You then backward bend the thoracic spine, which pushes this point downward. This movement increases the sternoumbilical distance and forces the stomach to enlarge its vertical diameter; fixing this point thus pulls the cardia downward. To increase the effect, add left rotation and right sidebending — the left hypochondrium will be stretched even more *(Illustration 5-12).* Repeat this technique slowly and rhythmically until the tissues relax. For best results, work on a few points during a single session.

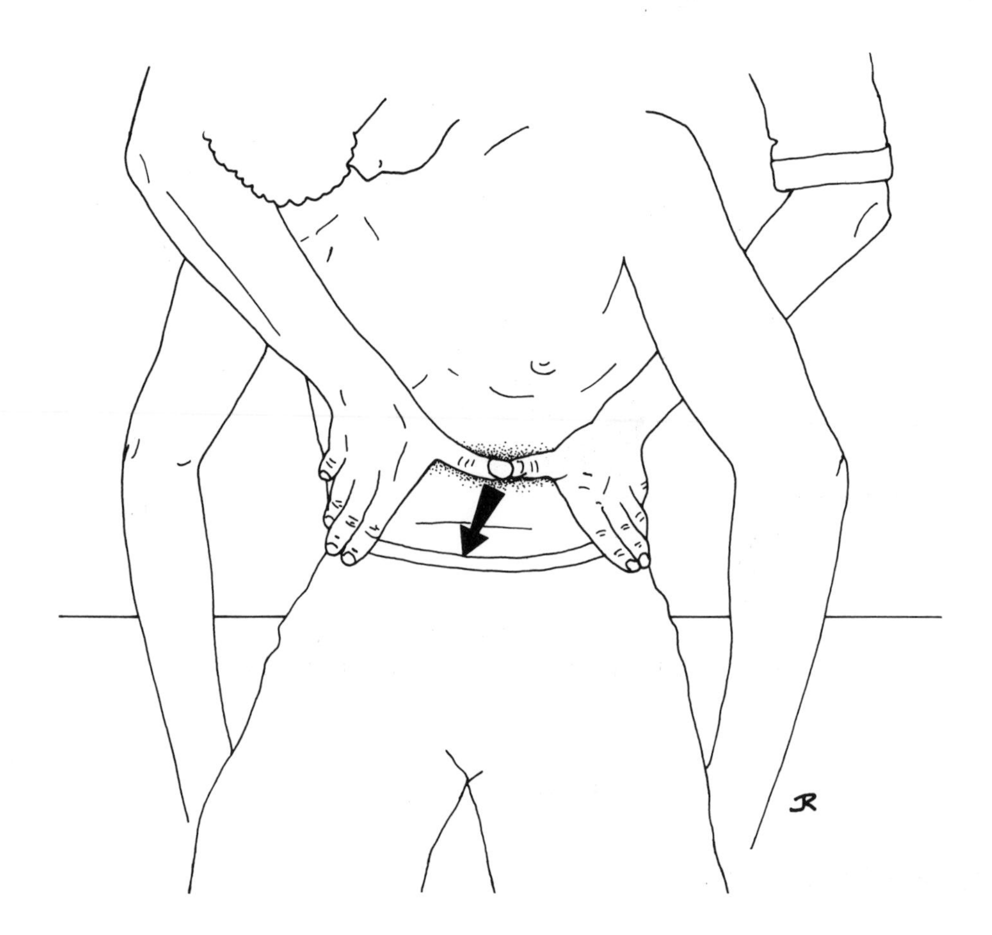

Illustration 5-12
Manipulation of the Gastroesophaegeal Junction

Gastric Ptosis

This technique consists of pushing the gastric fundus and a part of the greater curvature posterosuperiorly, with the patient in the seated position and the fingers under the costal edge on the midclavicular-umbilical line. Gradually, increase the patient's kyphosis in order to push the hand as high up as possible (but without causing pain, as an abdominal contraction will push away the fingers) and finish with a left rotation of the trunk to increase the distance between the gastric fundus and the pyloric antrum.

At the end of the movement, the patient is righted progressively by pulling the stomach upward and toward the left with the fingers. The sternoumbilical distance increases and the pyloric antrum is forced to move upward *(Illustration 5-13)*. We have checked this technique several times by gastric fluoroscopy and seen the pyloric antrum move upward by as much as 5cm.

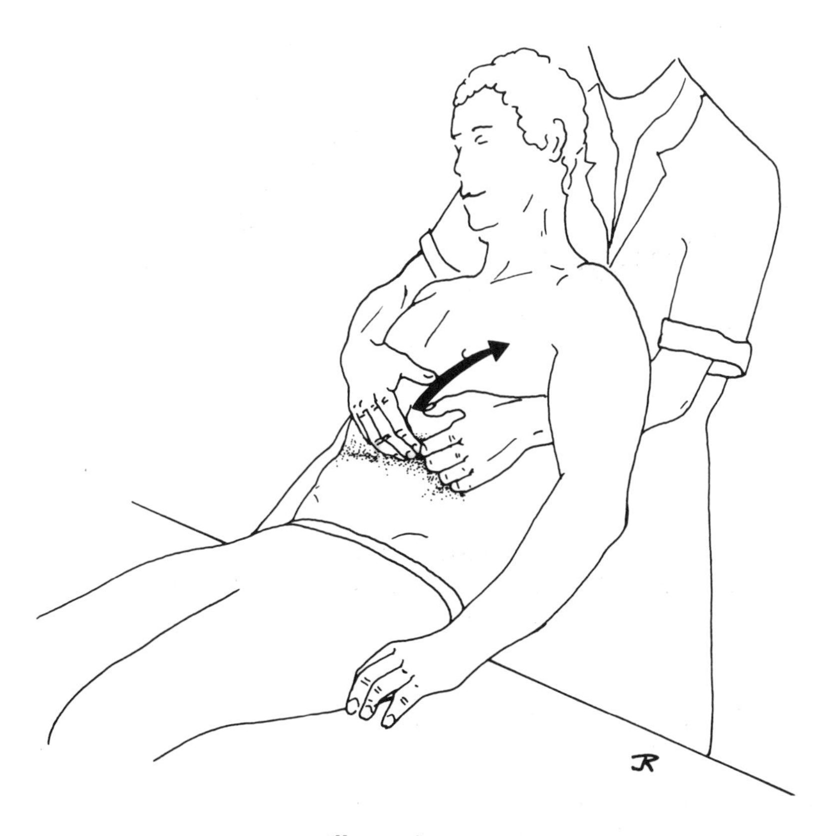

Illustration 5-13
Manipulation for a Gastric Ptosis

We recommend using the reverse Trendelenburg position because of its great efficiency in cases of gastric ptosis and even advise our patients to spend time using it at home. The technique enables them to feel the movement well. There are two methods:

- the patient rests the nape of the neck and the upper thorax on the ground, his pelvis on your thigh and his feet on the table *(Illustration 5-14)*
- you sit on a stool; the patient rests the back of his neck and upper thorax on your thighs and his feet and pelvis on the table *(Illustration 5-15)*.

The patient will be surprised by this position at the beginning but will soon feel secure. This position enables you to sidebend the patient with your two free hands well into the left subcostal region. In this position, the stomach is manipulated in the same direction

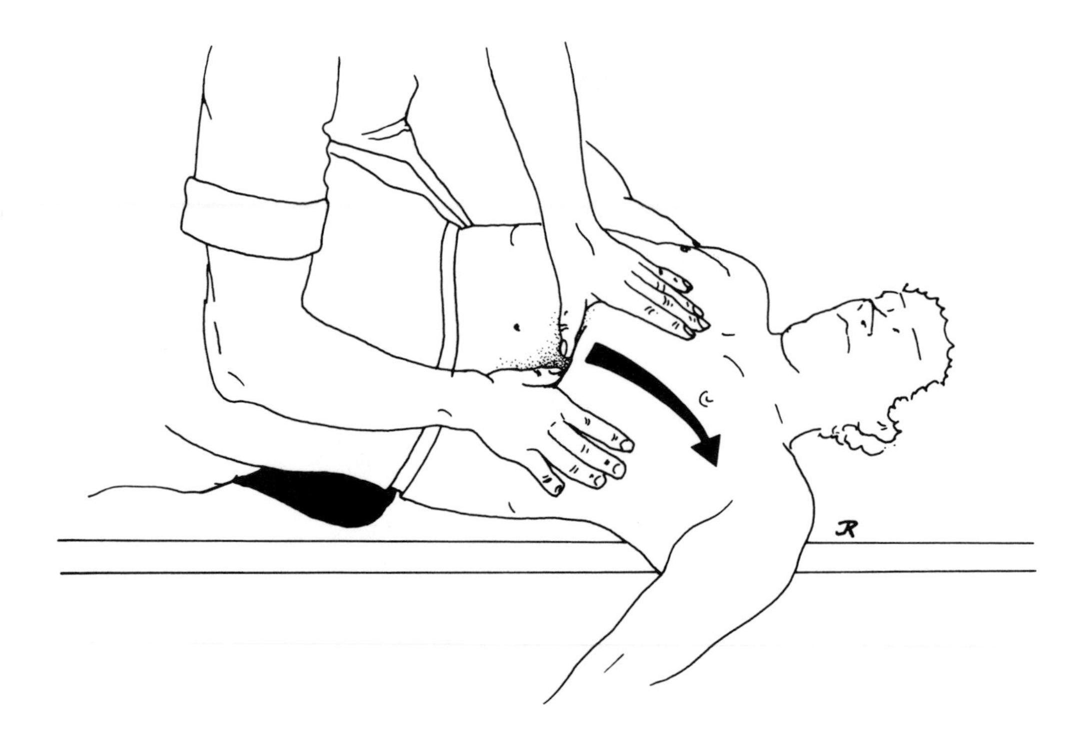

Illustration 5-14
Stomach Manipulation in the Reverse Trendelenburg Position — First Method

as in the seated position, with much of the work done by gravity. We highly recommend this excellent technique.

These techniques may move the stomach upward but not actually hold it up in place. We will stress here again that *motion* is much more important than position. In any case, it is important that the stomach moves again in all directions, which can be accomplished by these techniques. If you are not convinced about the efficacy of your actions, listen to the area of the pylorus with a stethoscope while performing the manipulations and you will hear the typical noises of gastric transit.

Air Pocket

Use the kyphosed seated position, with the fingers against the gastric fundus. To treat this area, mobilize the mass of air with local movement to break it up and push it superomedially toward the cardia. Adding left sidebending of the trunk will help. Often the patient will belch after this movement, which proves its efficacy. This technique has an antispasmodic effect. A large air pocket in the stomach is due to a spasm of the stomach muscles, which interferes with proper mixing of gastric contents. This technique decreases gastric transit time and thus avoids stasis phenomena which generate gastritis.

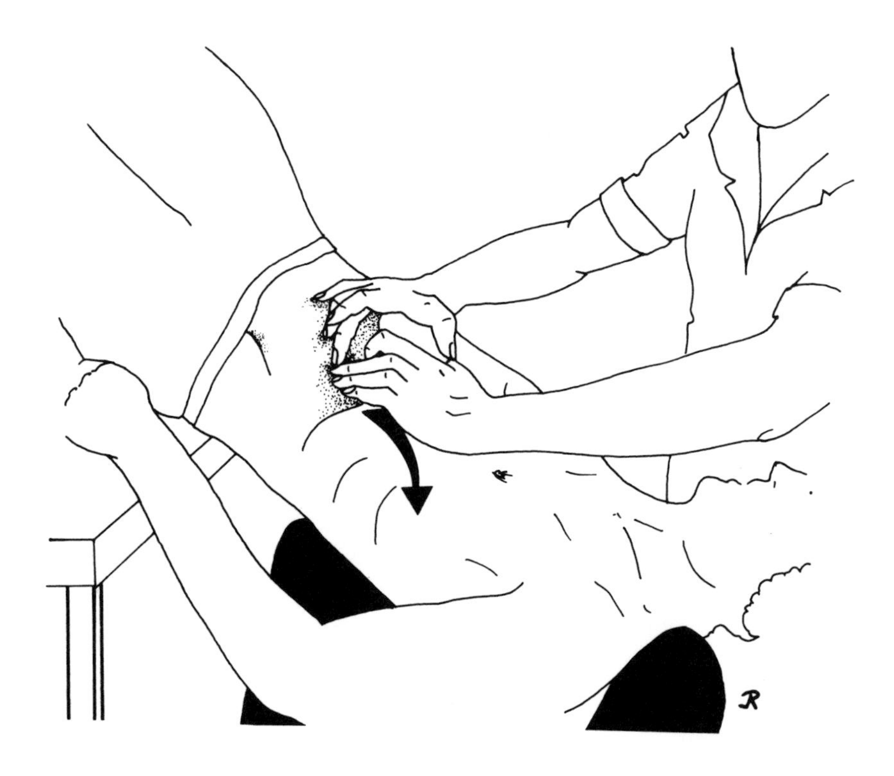

Illustration 5-15
Stomach Manipulation in the Reverse Trendelenburg Position — Second Method

Pylorus

This is not easy to find unless in spasm. Its position varies (as we have noted several times), but it is usually found approximately 6-7cm above the umbilicus. Place the patient in a supine position, legs bent, with the head on a cushion. Apply the heel of your hand against the spot where you believe the pylorus to be. Use the thenar or hypothenar aspect of your hand, whichever is more sensitive. Push posteriorly as deeply as possible without causing pain; you should feel a relatively firm area. If it is the pylorus, you will feel a rotational motion. If you do not feel this, palpate again until you find the pylorus. Let up just a bit on the pressure and then follow the clockwise and counterclockwise rotation 4-6 times until the tension releases. Repeat up to 10-12 cycles until the relaxation of the sphincter is complete. This may be accompanied by a "whooshing" sound. Normally the primary motion of the sphincter is a clockwise rotation. If the technique is difficult to accomplish, first prepare the area by using a translatory movement back and forth from left to right. If the pylorus was involved, the patient quickly feels the pain fade away.

It has been our experience that if one of the major sphincter-like areas of the digestive tract is in spasm, the rest are likely to be too tense also. Therefore, if the pylorus needs to be treated, you should also evaluate the sphincter of Oddi, duodenojejunal junction and ileocecal junction. The techniques for locating and treating the sphincter are all the same; only the location changes.

COMBINED TECHNIQUES

Among the many possible techniques, we prefer that which combines a direct movement (short lever arm) with an indirect movement (long lever arm) by sidebending of the lower extremities. With the patient in a supine position, arms apart, place your right forearm under the knees in order to lift up the legs. Place your left hand at the level of the greater curvature, or behind R9/10, which will be pushed posterosuperiorly. This is a direct rotational manipulation of the stomach toward the left. The combination of the movements of the lower limbs on the trunk will accentuate this rotation *(Illustration 5-16)*.

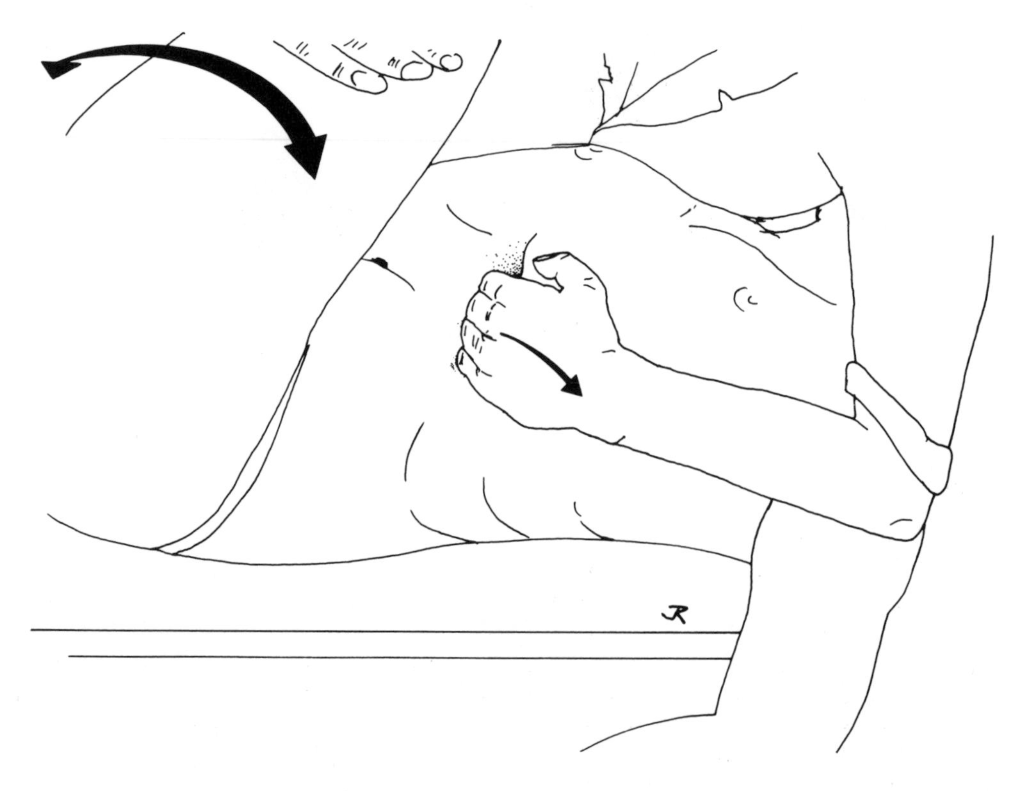

Illustration 5-16
Combined Manipulation of the Stomach

A combined technique that is useful for hiatus hernias is via the neck and head. In addition to the direct techniques described above, have the patient grasp his hands behind the neck while you hold on to his elbows, using them as a lever to rotate the body to the right. At the same time extend the body and pull it up in order to accentuate the stretching of the fibers of the stomach. Do this rhythmically until you feel a release. Be sure to respect the slightly oblique course of the esophagus during this movement.

INDUCTION TECHNIQUES

These are performed with the patient in a supine position and your hand in the same position as for the motility tests (page 125). If you find that the stomach goes into expir more easily and fully, then you will follow and encourage the expir movement. This is performed as if you wished to push the gastric fundus inferomedially and the pyloric antrum superolaterally. The hand performs a clockwise rotation in the frontal plane. At the end of the movement, the lesser curvature (under the index finger) will move posterosuperiorly. The right hand should be flat in order to create a suction effect with the skin, which will make this movement easier to feel. At the end of the movement, the ulnar aspect has the sense of moving away from the skin *(Illustration 5-17).* As usual, induction is accomplished by encouraging the predominant motion and passively following the other part of the cycle until a release (with or without a still point) is felt.

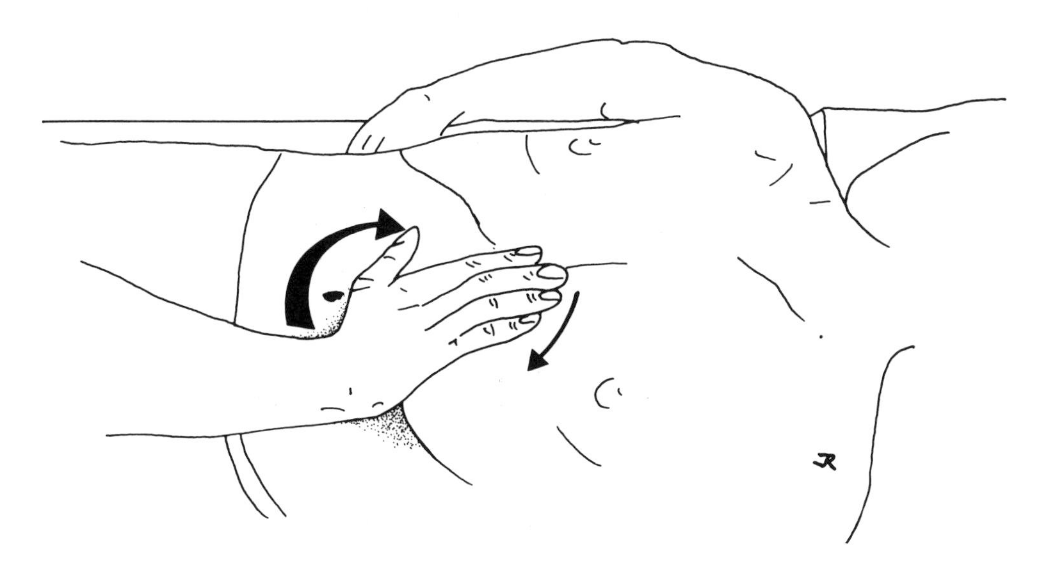

Illustration 5-17
Stomach Induction — Supine Position

You can also induce the cardiac zone alone in the supine position (or, if preferred, the seated position). Place one hand at the level of T11 (zone of posterior projection of the cardia) and the other on the xiphoid. If there is a problem with the cardia or esophageal junction, the posterior hand will be pulled anteriorly. With the two hands working in concert and going in opposite directions, perform a movement of the anterior hand to the upper right and then lower left (opposite for the posterior hand) *(Illustration 5-18).* Although the hands should be subtly active during this induction, for this technique to work best they follow and encourage motion in the easiest direction. If there is an anterior pull at the start of the technique, it should release by the end. We like to finish treatments of the stomach with this technique.

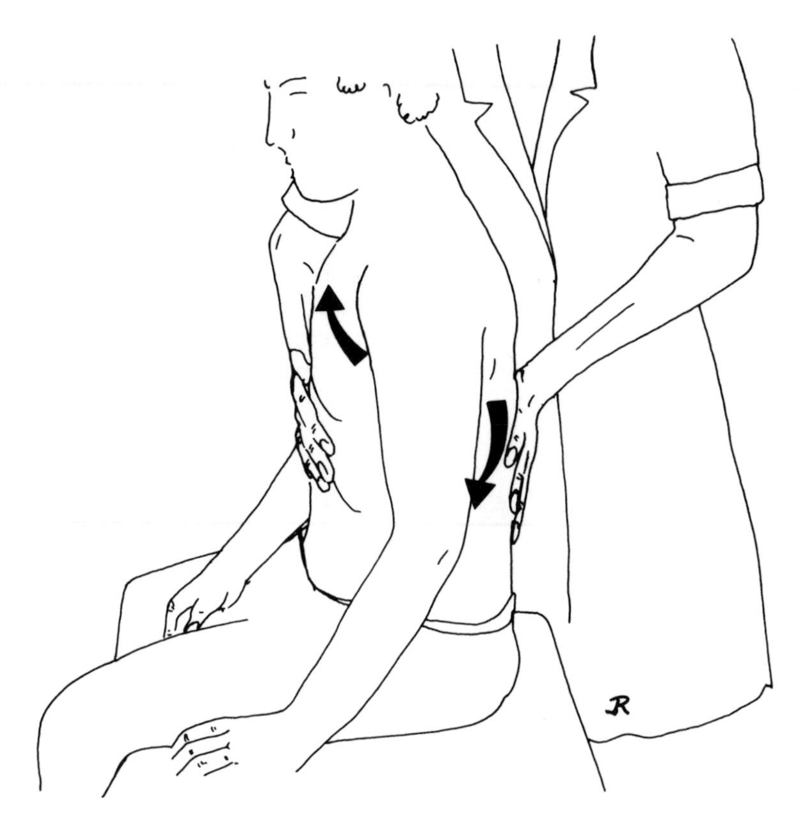

Illustration 5-18
Anterio-Posterior Induction of the Cardia — Seated Position

Adjunctive Considerations

ASSOCIATED OSSEOUS RESTRICTIONS

We hesitate to mention the standard solution that stomach dysfunctions are related to restrictions of T6, but sometimes visceral manipulations are not enough and we need help from the reflexogenic costovertebral keyboard. T6 does correspond to the stomach's dermatome, T11 to the topographical projection of the hiatal region and left L1 to the left diaphragmatic crus, which has a direct relationship with the cardia. Very often, in cases of gastralgia, ulcers and hiatus hernias, one finds an L1 restriction. Manipulation of L1 has an immediate effect on the diaphragm which, in relaxing, frees the tensions on the esophageal sphincter which thus functions better. One could almost say the practitioner is lucky if there is a restriction. T6-11 restrictions are not proof of a gastric problem, but will almost always result from restrictions in this region. There are sometimes restrictions further away in the cervicals (vagus) and in the left sacroiliac joint (related to L1).

RECOMMENDATIONS

The stomach takes two hours to empty and therefore the patient should avoid eating for two or three hours before treatment. In cases of hiatus hernia or other problems involving the gastroesophageal junction, diet is important. Foods that dilate the sphincter (e.g., oranges, wine, coffee, chocolate) should be avoided. In case of gastric ptosis, advise the patient to eat and drink in small amounts so that a great volume does not suddenly fill up the stomach. It is best if the patient drinks warm fluids, slowly (page 107). Advise against belts or trousers that are too tight, as well as activities with the arms raised and head extended. Have patients with ptosis rest in the reverse Trendelenburg position just before the evening meal, to accentuate the diaphragmatic respiration. The patient can pull the stomach upward himself. For obvious reasons, this position is inappropriate in cases of gastric reflux or hiatus hernia.

Chapter Six: The Small Intestine

Table of Contents

CHAPTER SIX

The Small Intestine

Despite their functional differences, one rarely treats the stomach without also manipulating the duodenum. Duodenal pathology is very frequent — who hasn't heard of a duodenal ulcer? This part of the small intestine is relatively constant in position and is an excellent starting point for visceral manipulation. Every time it is affected, it is painful, fixed and in spasm, the latter being easily perceived. When describing a viscerospasm, we usually refer to the duodenum. This condition is often found in thin, pale and asthenic subjects who are typical ulcer candidates. In contrast to the duodenum, the jejunoileum is a mobile mass which is more difficult to find and can cause real positional problems. Torsion of the loops of the jejunoileum are well known in surgery; less serious ones are our province. They do not imply a need for surgical intervention, but do provoke significant problems.

Anatomy

The duodenum is a continuation of the pylorus, from which it is separated by the pyloric groove. It ends at the duodenojejunal flexure, at which point it becomes free of the peritoneum. Its shape is roughly like an incomplete circle, usually divided into four portions (superior, descending, inferior and ascending). For convenience, we shall refer to these four portions as D1, D2, D3 and D4 respectively. The duodenum is found posteriorly in the abdominal cavity, usually on the right of the vertebral column in the supine position and during exhalation. It goes from T12 to L4, from the right subcostal region to the hypogastric region.

The jejunoileum, which continues from the duodenojejunal flexure to the ileocecal junction, is arranged in a series of loops attached to the posterior abdominal wall and covered by peritoneum. These loops are loosely attached by mesentery to the posterior wall and have considerable freedom of movement. In general, the jejunum (the proximal two fifths) occupies the umbilical region, while the ileum (the distal three fifths) occupies the hypogastric and pelvic regions.

RELATIONSHIPS

Duodenum

The duodenum traverses the midportion of the abdominal cavity and comes in contact with most of the contents of the peritoneal space *(Illustration 6-1)*. It consists of the four portions mentioned above, plus the caruncles. The duodenum is a relatively fixed organ. The most mobile part is D1, which is entirely covered in peritoneum. Its position can vary by 4-5cm depending on respiration or the position of the subject.

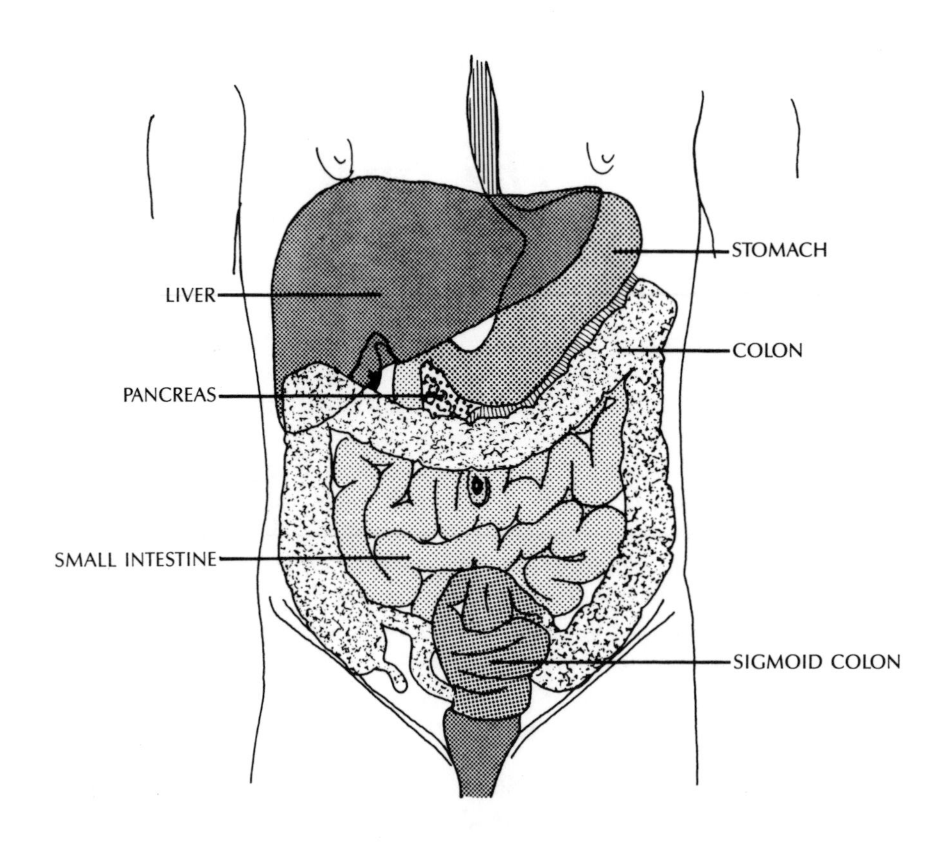

Illustration 6-1
Anatomical Relationships of the Duodenum

The superior duodenum (D1) relates to the liver. Obliquely directed superiorly, posteriorly and to the right, it goes under the liver at the level of the neck of the gall bladder. It is related to the body of L1 when seated, L2 when lying down. Anteriorly, it is related to the inferior side of the liver and the neck of the gall bladder, to which it is medial. The anterior side of the angle formed by D1 and D2 is in contact with the right lobe of the liver. The posterior side of D1 descends vertically and is connected by peritoneum to the posterior cavity of the omenta. The extraperitoneal segment, which is attached at the bottom to the neck of the pancreas, is connected with the trunk of the portal vein and the common bile duct.

The descending duodenum (D2) is situated on the right and in front of the vertebral column from L1 to L4. Anteriorly, it is covered by the peritoneum and connected to the right extremity of the transverse colon. The mesentery divides it into a supramesocolic portion covered by the right extremity of the transverse colon and a submesocolic portion connected to the floating loops of the jejunoileum. Posteriorly, D2 is covered by Treitz's muscle (suspensory muscle of the duodenum), and connected to the vena cava and the anteromedial part of the right kidney, to which it is often closely attached. It also interacts with the right renal hilum, ureter and spermatic artery. On the left, it connects with the head of the pancreas, its excretory canals and the common bile duct. On the right, in the supramesocolic part, it connects with the right lobe of the liver and lower down with the ascending colon.

The inferior or transverse portion of the duodenum (D3) runs from right to left, forming a concavity superiorly and posteriorly. It normally interacts with the body of L4 (which entails a risk of being crushed by a direct contusion). Anteriorly, it is connected to the root of the mesentery (which crosses it obliquely), the mesenteric vessels and the loops of the jejunoileum. Posteriorly, via Treitz's muscle, it rests on the psoas muscle, inferior vena cava and aorta. It is attached to the head of the pancreas superiorly, and is adjacent to the jejunoileum inferiorly.

The ascending duodenum (D4) goes obliquely upward to the duodenojejunal flexure (at the left lateral aspect of L1/2) and follows the left flank of the aorta and lumbar column. Anteriorly, it interacts with the pyloric antrum, transverse mesocolon and jejunoileum. Posteriorly, it relates to the psoas muscle and left renal vessels. On the right, it follows the aorta and the upper part of the mesentery.

There are two caruncles. The large caruncle is located on the posteromedial side of D2; in many cases it is penetrated by the ampulla of Vater or opens into Santorini's duct and the duct of Wirsung. The small caruncle is located 3cm further downward and opens into Santorini's duct — an inconsistent orifice which is not always present.

Jejunoileum

The duodenojejunal flexure is found above the transverse mesocolon to the left of L2 and medial to the descending colon. It is in contact with the left crus of the diaphragm. The flexure is suspended by Treitz's muscle. Its position on the left is directly opposite and the mirror image of the sphincter of Oddi's position on the right.

The jejunoileum itself measures 6.5m and makes 15-16 U-shaped intestinal loops in a regular order, divided into two groups: the upper left which are arranged horizontally and the lower right which are arranged vertically. These loops form a mass which fills the abdominal cavity, more on the left than on the right. On the left, they cover the descending colon, whereas on the right they leave the ascending colon free. Posteriorly, they relate to the posterior abdominal wall and to the retroperitoneal organs (vessels, submesocolic part of the duodenum, kidneys, ureters, ascending and descending colons). Anteriorly, they relate to the greater omentum (which covers the entire jejunoileum), superiorly with the transverse colon and its mesocolon and inferiorly with the organs of the pelvis, particularly the bladder.

VISCERAL ARTICULATIONS

The duodenum is held up by the peritoneum, which attaches the supramesocolic portion to the inferior parts of the liver, bile duct and posterior abdominal wall. On this fairly mobile portion, the peritoneum surrounds the anterior and posterior sides of the duodenum in the same way as it does the stomach. Superiorly, the two layers contribute to the formation of the lesser omentum.

D1 continues superolaterally with the peritoneum which covers the inside of the right kidney and then with the part which covers the anterior pancreas. At the bottom, it forms the superior layer of the transverse mesocolon.

Posteriorly, the supramesocolic portion of the duodenum is fairly well attached to the anterior vertebral column. This is the most fixed part of the small intestine; the rest is relatively "floating."

On the right, the submesocolic duodenum consists of D2 (the inferior part) and D3. Above, the peritoneum follows the anterior side of the pancreatic head and bends backward to form the inferior layer of the transverse mesocolon.

The left submesocolic duodenum consists of D4 and the duodenojejunal flexure, of which only the anterior portion is covered. Superiorly, the peritoneum joins the transverse mesocolon; inferiorly, the left psoas muscle; on the left, the left kidney and descending mesocolon.

The peritoneum covering D4 and the duodenojejunal flexure forms semilunar folds called "fossae," which carry the risk of strangulation of the small intestine. Three in number, they are the result of an incomplete attachment of the duodenum to the posterior parietal peritoneum. The most constant are called the superior and inferior fossae.

Treitz's muscle, made up of smooth fibers, connects the duodenojejunal flexure to the left crus of the diaphragm. We are not sure of its function — it might play a sphincter-type role for the evacuation of the duodenum. This is a zone which often hardens with palpation and has a sphincter-like rotation. It is possible that the muscle has a role in this hardening and rotation.

The mesentery is a large peritoneal fold joining the jejunoileum to the posterior abdominal wall. It comprises its own ligament with two edges and two sides. The posterior edge, 16-18cm long, joins the posterior peritoneum to the small intestine. It arises from the cecum and rests on the internal side of D4, the abdominal aorta, vena cava and primary iliac vessels. The anterior edge measures about 6m and continues with the peritoneal layer which covers the small intestine. Because of the size of this edge, the jejunoileum, held by the posterior edge, is totally mobile. The mesentery has two layers which enclose the vessels of the small intestine — this explains its importance in manipulations.

Intracavitary Pressure and Turgor

These are the most important elements in maintaining the jejunoileum. There is considerable pressure inside the small intestine due to the gas it contains; this is one of the important regulators of intracavitary pressure. During a laparotomy, it is remarkable to see the loops of the jejunoileum which try to escape from the abdominal cavity — proof of the important maintenance role of abdominal pressure. The mesentery also plays an important support role; other elements are more involved in the visceral cohesion factor.

Sliding Surfaces

Sliding surfaces are found between the small intestine and numerous other organs, either directly or via the peritoneum. The most important are the liver, gall bladder, common bile duct, pancreas, transverse colon, ascending and descending colons, kidneys, psoas muscles and stomach. The jejunoileum can also articulate with the urogenital organs in the case of a ptosis.

TOPOGRAPHICAL ANATOMY

In general, the pylorus is found 6-7cm above and slightly to the left of the umbilicus when the stomach is empty, and 3-4cm to the right of the umbilicus when the stomach is full. However, its position may vary considerably (page 116).

D1 is to the right of L1-L4 *(Illustration 6-2)*. It is subhepatic and deeply placed in its superior part (its hepatic impression is found slightly anteromedial to the right kidney). To reach it, use the subcostal technique with the patient in a kyphosed seated position, placing the fingers inferomedial to the bile duct (or just a little above the pylorus when it is on the right side).

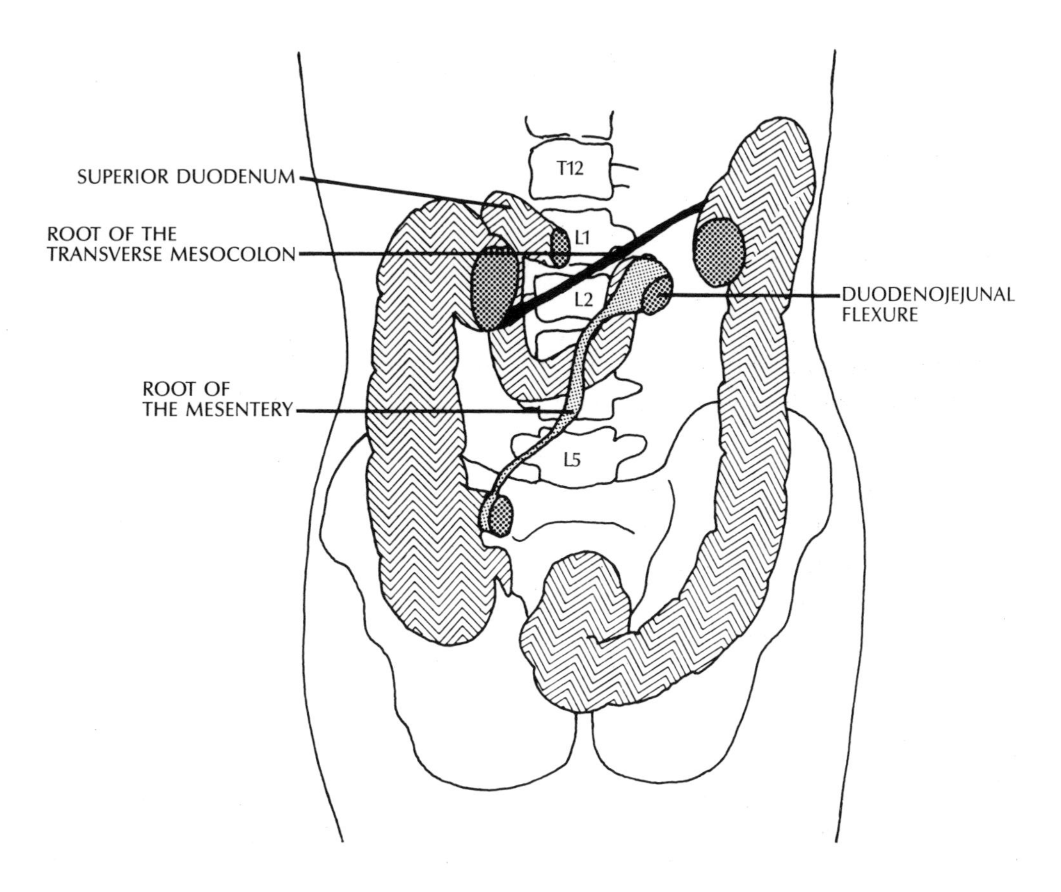

Illustration 6-2
Topographical Anatomy of the Duodenum

To find D2, first find the ascending colon. On its medial side, at the level of the umbilicus, another pipe-like structure can be felt; this is D2. It becomes more difficult to feel as you go superiorly, but sometimes D1 can be palpated in this manner if the patient is very thin.

The sphincter of Oddi is found on a line between either the right midclavicular point or the right nipple and the umbilicus, 2-3 fingers' width above and to the right of the umbilicus, at a point roughly opposite the duodenojejunal flexure. It is located in the posteromedial portion of D2 and can often be reached across the anterior wall. One reaches it via the greater omentum, transverse colon or jejunoileum, depending upon their position. It is often sensitive to pressure.

D3 forms a bridge over L4, while D4 is located to the left of L2-4. The duodenojejunal flexure is found on the upper left edge of L2/3. It is 2-3 fingers' width above the umbilicus, on a line between the umbilicus and either the left midclavicular point or the left nipple. One reaches it in the same way as the sphincter of Oddi.

The jejunoileum is situated inferior to the umbilicus behind the greater omentum. It is primarily in the left part of the abdomen, but does go over the lower right quadrant where it meets the cecum (page 167).

Physiologic Motion

MOBILITY

It is not possible to be as precise here as in describing liver and stomach mobility. This is perhaps due to the relative remoteness of the duodenum from the surface and its length. In manipulation, the pyloric region, sphincter of Oddi and duodenojejunal flexure are the most important areas.

The duodenojejunal region is relatively fixed and often used as a reference mark in radiology. With diaphragmatic movement, it is mostly D1 and D2 which become mobile. During inhalation D1, interdependently with the liver, moves inferomedially. The duodenum, due to its roughly circular shape, folds up upon itself, closing the angles, and D1 is lowered by the width of at least half a vertebra. The direction of movement follows that of the falciform ligament; D1 and D4 move together. There is a sagittal movement but it is very slight and has little importance in the realm of manipulation.

Mobility of the jejunoileum is much less influenced by diaphragmatic movement, and we do not feel that it is useful to describe it here. Mobilization can really only take place through working on its attachments, primarily the root of the mesentery. It is impossible to precisely describe the total path of jejunoileal mobility, because of its length.

MOTILITY

Expir is the part of the visceral motility cycle that takes an organ toward the median axis of the body. The duodenojejunal flexure is relatively fixed and therefore D4 does not move significantly during the motility cycle. As a result, expir for the duodenum is a circular clockwise motion during which the other three portions draw closer to the median axis of the body, i.e., the vertebral column *(Illustration 6-3)*.

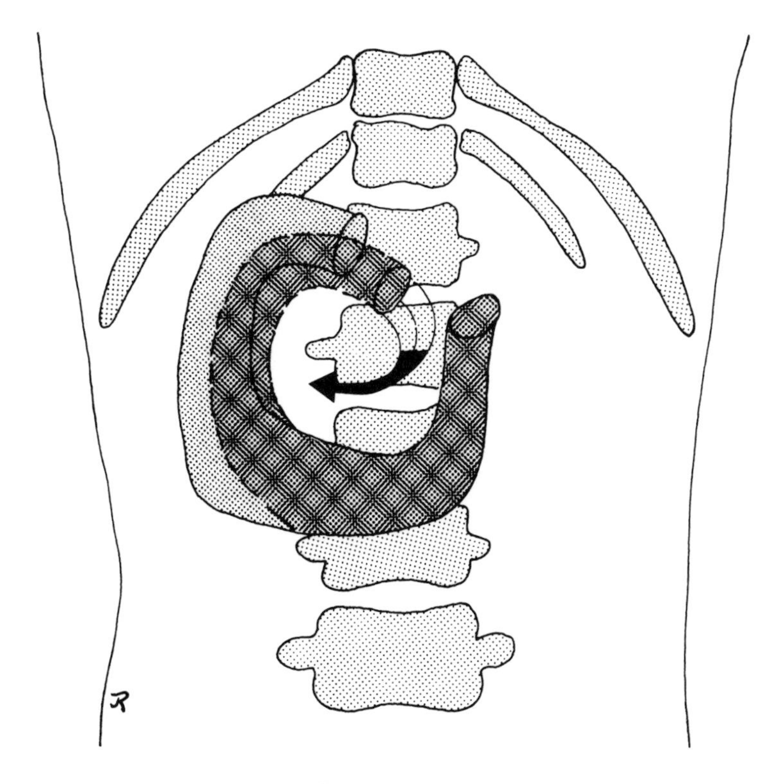

Illustration 6-3
Motility of the Duodenum — Expir

At the end of its embryological development, the inferior extremity of the stomach rotates toward the right, taking with it the duodenum in a right rotation around a vertical axis. It also swings in a clockwise rotation around a sagittal axis, the superior extremity of the stomach moving to the left, while the duodenum moves slightly upward and to the right. This is another example of visceral motility recreating the motion of embryogenesis.

For most sphincters, including the sphincter of Oddi, inspir involves a clockwise rotation and expir a counterclockwise rotation.

Motility of the jejunoileum is again inspired by the movement seen in formation of the mesenteries, following a clockwise direction during the expir phase. The ascending branch goes from left to right, and thus the descending branch moves from right to left, which places it behind the ascending branch *(Illustration 6-4)*. The direction for a correctional movement of the jejunoileum is from left to right and bottom to top, following a clockwise direction.

Indications for Visceral Evaluation

All inflammatory symptoms, whether septic or not, present a risk of adhesion or a significant decrease in mobility. For the duodenum, the most important indication is

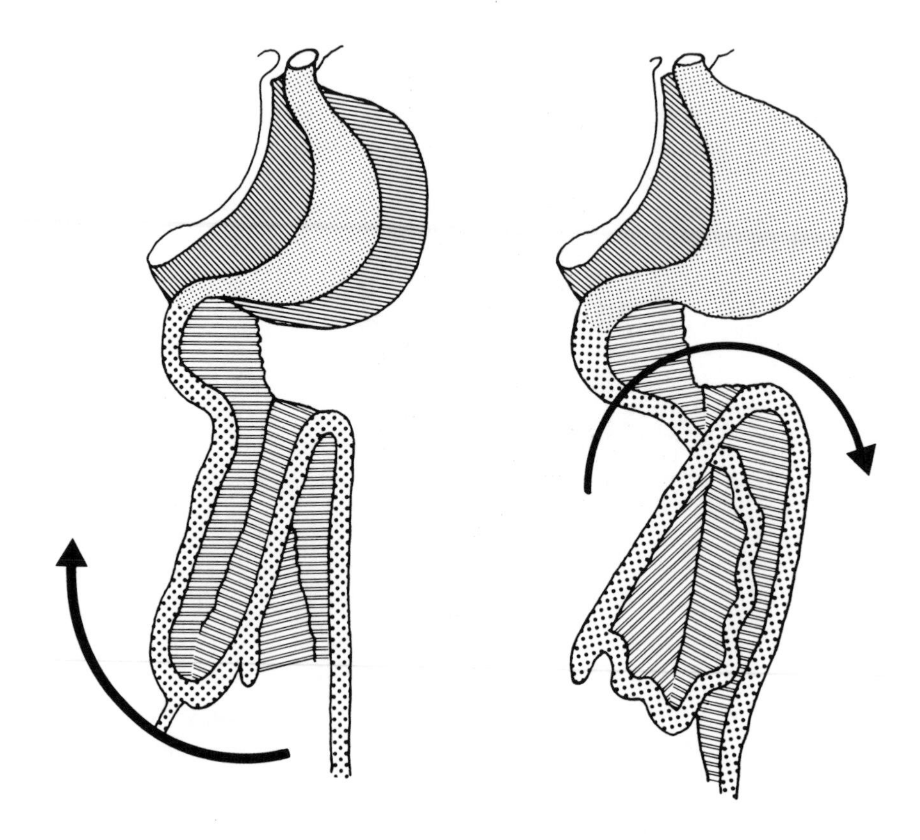

Illustration 6-4
Intestinal Twisting During Embryogenesis

an ulcer and its sequelae associated with visceral spasm, which then fixes all surrounding regions. Gastric, biliary and pancreatic transit diminish and the individual cannot eliminate the toxins, which collect in the intestinal tract. The duodenum goes easily into spasm; most treatment is therefore antispasmodic in function.

There are often indications for visceral manipulation following surgery of the small intestine. Mobility of this organ is essential for correct digestion. The smallest adhesion or too tight a bend can cause an obstacle to intestinal transit; the 6.5 meters must slide consistently and harmoniously. Serous fluid is necessary for movement of the small intestine, and it is here that one most often finds cases of non-serosity in the abdominal viscera. An appendectomy can frequently play a large part in restrictions here. It may fix the root of the mesentery by drawing it downward and to the right, so much so that there is a pathological change of the jejunoileal axis. The patient in these cases often complains of postoperative constipation.

The small intestine has a complex circulatory, lymphatic and nervous system supply. Manipulation will increase the circulation of fluid and release spasms of the sphincters and intestinal wall. Because of its microvilli, the total surface of the small intestine is approximately 100m^2. This organ must be mobilized in cases of intestinal ptosis, a frequent illness after 50 years of age which often accompanies gastric ptosis. The principal symptoms are: a feeling of unease and pulling under the umbilicus, mostly 3-4 hours after a meal; difficulty in wearing belts and tight trousers; lower abdominal pain after standing for a long time; and breathing difficulty when standing which is improved upon lying down.

Evaluation

The patient history should clarify the nature of the symptomatology. It is important to check the type of pain, its frequency and duration, and its relation to food intake. Do be careful about ulcers and risks of perforation. Do not hesitate to ask the patient about the color of bowel movements. If there is any doubt, insist on radiography or endoscopy. With intestinal ptosis, biotype morphology is again important; a longiline asthenic type is the most likely to develop this condition. Watch people who present themselves with abdominal distention from ptosis of the small intestine — you will see that the distention generally begins just underneath the umbilicus, following an oblique line corresponding to the roots of the mesentery. Usually digestive distress related to the small intestine appears 3-4 hours after eating. For women, it is essential to find out if any of these problems began after pregnancy, which tends to stretch the whole system of visceral attachments because of edema of the tissues.

Palpation of the duodenum is difficult. We shall describe it in detail in the following section. Palpation of the jejunoileum is easier, and painful where there is a problem. To confirm a case of intestinal ptosis with a patient seated or standing, lift up the lower part of the jejunoileum and let it fall suddenly — if this technique is painful, it confirms the involvement of the jejunoileum in the patient's problems. Percussion of the abdomen should also reveal abnormal tympany in these cases.

MOBILITY TESTS

D1 is subhepatic and deeply placed in its superior part, and therefore its mobility is difficult to evaluate. When attempting to reach it, use the subcostal technique with the patient in a kyphosed seated position, placing the fingers inferomedial to the bile duct (or just a little above the pylorus when it is on the right side) and sidebend the patient to the left. Push posterosuperiorly; you may be able to feel a tubular structure, and if there is excessive resistance or tightness here it may be a restriction of D1. This test is fairly difficult to perform and interpret.

An alternative is to test D1 via the liver by lifting that organ in the kyphosed seated position (page 92). It is not possible to differentiate between restrictions of the liver and those of D1 by means of this test.

The most important part of the mobility test is direct, local palpation to appreciate the state of induration and relative motion of the sphincter of Oddi and the duodenojejunal flexure. This is accomplished by placing the thenar or hypothenar aspect of the heel of the hand (or the pads of a few fingers) directly over the area to be tested. For the sphincter of Oddi and duodenojejunal flexure, these areas are 2-3 fingers' width above the umbilicus, and on the right and left midclavicularumbilical lines respectively. Press down as deeply as possible without causing pain and then let up slightly. If there is no problem with the area, you will feel an easy rotational motion that is more active in a clockwise direction. If you do not feel any rotation, you are probably pressing in the wrong place.

In order to feel the root of the mesentery, have the patient lie supine or in the left lateral decubitus position with hips and knees flexed. The root lies on a line between the duodenojejunal flexure and the ileocecal junction. Place your finger parallel and 3-4 cm inferior to this line. Press posteriorly with your fingers taking a hook-like shape and then pull them toward the line. The root is palpable as a firm, thin structure. Palpate for areas of increased tension.

In the case of the jejunoileum, it is necessary to take different pressure points on the abdominal wall following the arc of a circle bounded by the cecum, descending colon and superior bladder. The test consists of pulling these different points toward the root of the mesentery to appreciate the resistance and elasticity of the jejunoileum and its connections; a fixed zone will be painful and necessitate greater traction.

MOTILITY TESTS

These are done in the supine position, with your hand flat on the patient's abdomen. There are two different methods.

For the duodenum, place your right hand slightly above the umbilicus, the ulnar edge going slightly to the right of the median axis of the body and the stretched-out thumb over the area of the sphincter of Oddi. During expir, the hand makes a clockwise rotation *(Illustration 6-5)*.

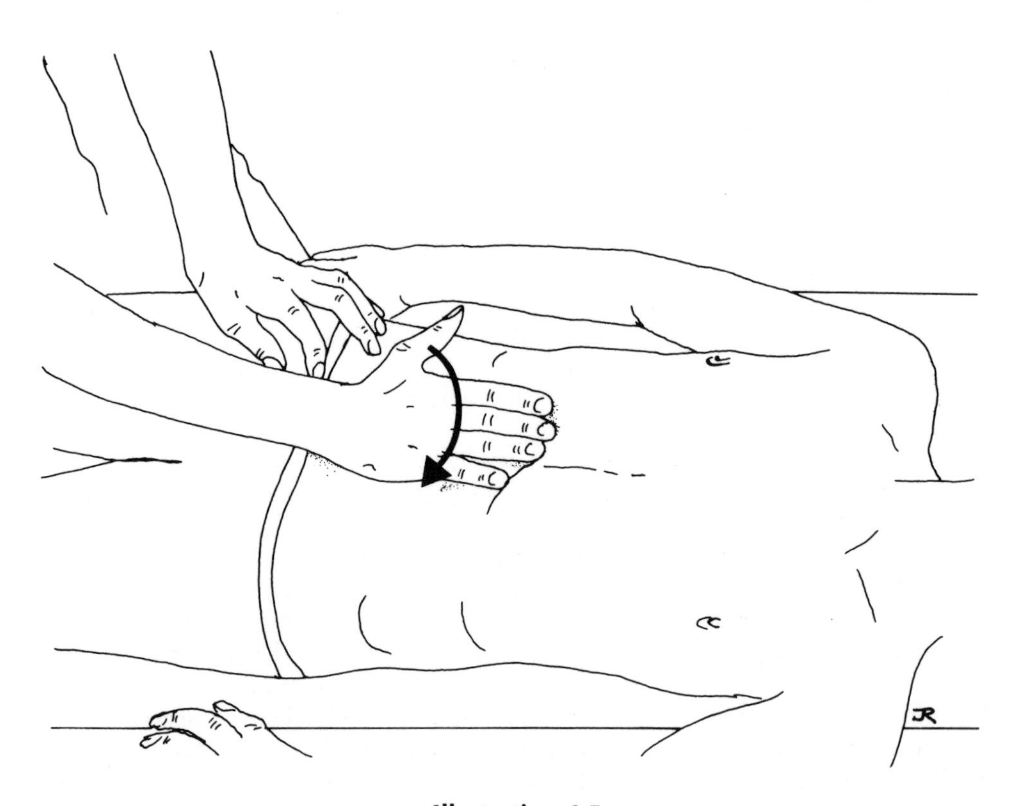

Illustration 6-5
Motility Test of the Duodenum

For the jejunoileum, use both hands — one following the direction of the upper transverse portion on the left and the other the vertical direction of the lower portion on the right. During expir, these two hands move inferomedially and superomedially respectively; i.e., they approach each other and the median axis with a clockwise motion *(Illustration 6-6).*

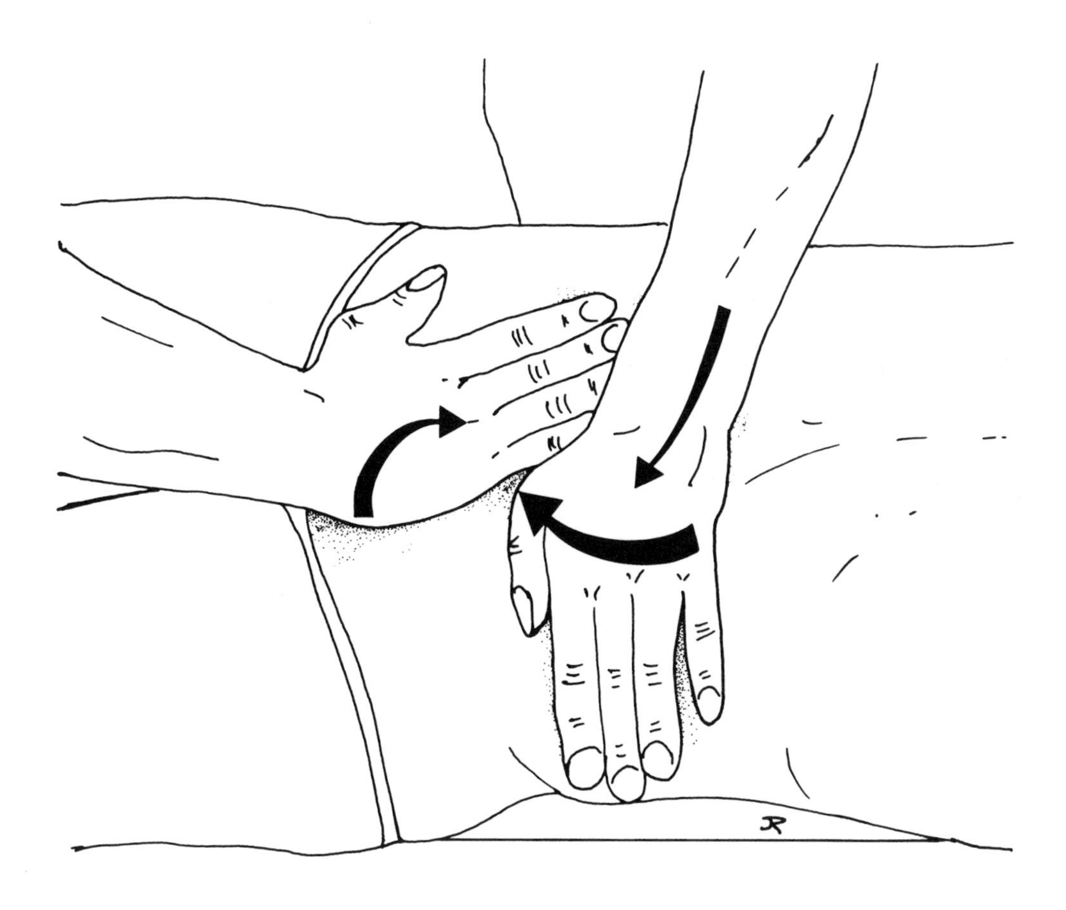

Illustration 6-6
Motility Test of the Jejunoileum

Restrictions

The most common restriction of the duodenum is that related to spasmodic factors (viscerospasm) and to the sequelae of an ulcer which fixes the duodenum posterolaterally. All surgical interventions on the small intestine can produce restrictions and all laparotomy patients must be tested. Restrictions on the jejunoileum tend to pull it downward and to the left in the case of sigmoiditis, and downward and to the right after appendicitis or appendectomy.

Manipulations

DIRECT TECHNIQUES

Duodenum

It is difficult to be certain of actually touching D1 as it is so far posterior. If this area requires mobilization, put your fingers just a little above the pylorus in the position described on page 149, and push them posterosuperiorly in the direction of the liver. Repeat this procedure rhythmically and with focused attention until a release is felt.

A method which is useful when you are unsure of the precise location of D1 is hepatic lifting, in the seated position with subcostal pressure (page 96). D1 is connected to the liver by the hepatoduodenal ligament; when the liver is mobilized upward, the duodenum follows, freeing restrictions.

With the patient in the right lateral decubitus position, you can try to mobilize the duodenum by placing your fingers on the medial edge of the ascending colon and lifting D2 and D3 upward toward the midline, away from the table, while simultaneously spreading the hands apart *(Illustration 6-7).* The spreading movement helps to break up local adhesions.

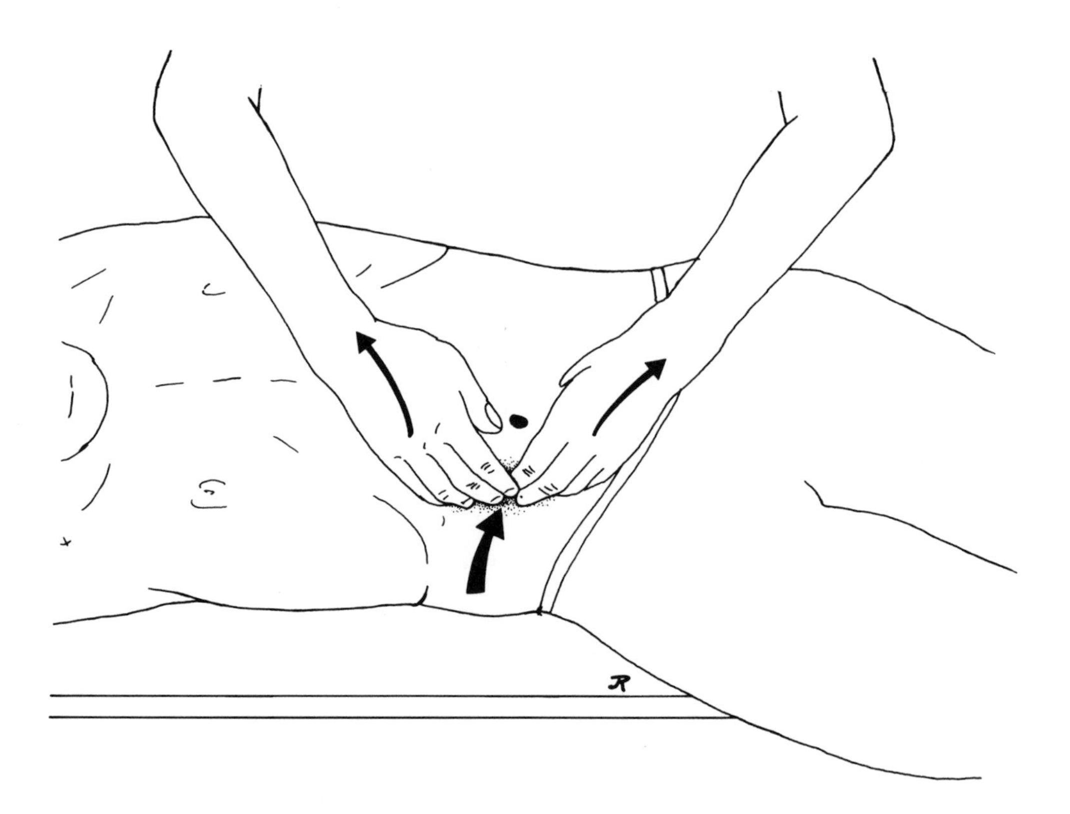

Illustration 6-7
Direct Manipulation of the Duodenum

A good way to mobilize the duodenojejunal flexure is via its attachment to L1. With the patient in the seated position, press on the abdomen anterior to the flexure until you can feel the flexure (not the left kidney). With your other hand, gently pinch the spinous process of L1 between your right thumb and index finger. Pull L1 to the right while pushing the flexure to the left. Repeat this gently and rhythmically until you feel a release, which should be clearly palpable in the lumbar region.

Jejunoileum

Imagine that the umbilicus is the central hub of a wheel and that the attachments of the jejunoileum are the spokes of this wheel (actually the wheel is a little to the left of the midline). Manipulations of the jejunoileum consist of drawing it (the arc of the wheel) superiorly and then medially in the direction of the umbilicus (except for the upper part, which is drawn medially and slightly inferiorly). Concentrate the mobilization over areas of tightness and repeat rhythmically until the areas of restriction release. This movement can be done in the seated, left lateral decubitus, or reverse Trendelenburg position — we recommend the latter as being the more effective position.

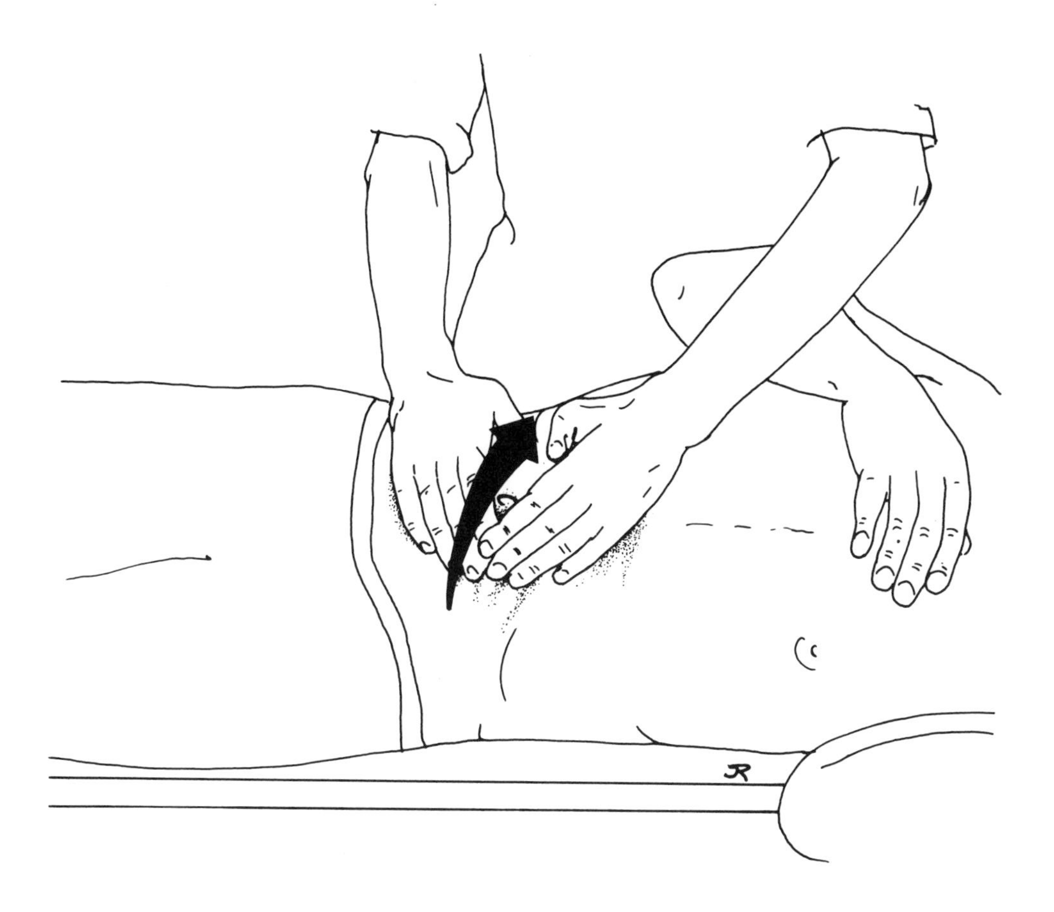

Illustration 6-8
Direct Manipulation of the Root of the Mesentery

The roots of the mesentery go from the ileocecal junction to the duodenojejunal flexure, forming an angle of 45 degrees with a transverse plane of the body. Place your hand on the line parallel to this line but toward the anterior superior iliac spine (ASIS). Push your hand first toward the back, then superomedially toward the umbilicus, in a gentle and rhythmic manner *(Illustration 6-8)*. If there is a restriction in the mesentery, you will feel your movements gradually freeing the attachments. This movement is usually done in the supine position but can also be done in the left lateral decubitus position, with the help of the finger pads. You can pull the mesenteric roots transversely by placing the thumbs or the pisiform in the middle of the line between the umbilicus and the ASIS and redirecting it toward the pubic symphysis, letting it come back and starting again — as if playing with the string of a bow. Be sure not to apply pressure to the aorta or femoral arteries.

Direct Pressure Technique

For the pylorus, sphincter of Oddi and duodenojejunal flexure, the hypothenar or thenar aspects (whichever is more sensitive) of the heel of the hand is used. Apply a progressive posteriorly directed pressure over the sphincter being treated until you have gone as far as possible without hurting the patient. Be careful not to confuse the duodenojejunal flexure with the left kidney. This is a common mistake and results from not looking for the rotational motion of the flexure. At this point, gently release the pressure and encourage the clockwise rotational motion rhythmically until you perceive a release (usually 5-10 cycles); then encourage the counterclockwise motion for a similar period of time until another release is felt. When the technique has been successfully completed, you should feel a release, and the motion should settle into a gentle rotation in which the clockwise phase is the most active. This technique is often followed by induction (see following section).

INDUCTION TECHNIQUES

The methods utilizing direct pressure on the sphincters and those of induction are quite similar. It is up to you to make the choice. With induction, however, you use less force and the direction and rhythm of the movements are determined by the inherent motion of the tissues. You follow and encourage whichever motion (clockwise or counterclockwise) is easier and with a greater amplitude until the tissues release. During the induction, you may or may not pass through a still point. Perhaps of all the areas of the body, the difference between the force used in a direct technique and that used in induction is smallest here. In practice, we usually start with direct pressure techniques and end with induction.

Duodenum

Induction of the duodenum begins with the hand in the same position as for the motility test (page 150). Follow the motion in inspir (counterclockwise) and expir (clockwise) phases of the motility cycle. Induction consists of following the motion, emphasizing the predominant phase until a release is perceived, with or without first going through a still point. If the motion is difficult to appreciate, try pushing D2 to the left and right with a translatory force several times to stimulate it before performing induction.

Sphincter of Oddi

The aim of induction here is to increase the transit of materials in the area, and the release of spasms in the muscular fibers. With the patient in the supine position, press the heel of your hand on the right midclavicular-umbilical line approximately 2-3 fingers' width above the umbilicus. Between your hand and the sphincter are the anterior abdominal wall, greater omentum, transverse colon and small intestine (depending upon their position) and D2; the pancreas is medial and the right kidney is lateral. Follow the clockwise/counterclockwise rotations carefully and gently until movement ceases *(Illustration 6-9)*. This is usually followed by a still point of varying length and then a release. You often hear a characteristic "drainage" noise. The sphincter of Oddi may be found lower down in cases of gastric ptosis. It is almost always affected by duodenitis and duodenal ulcers.

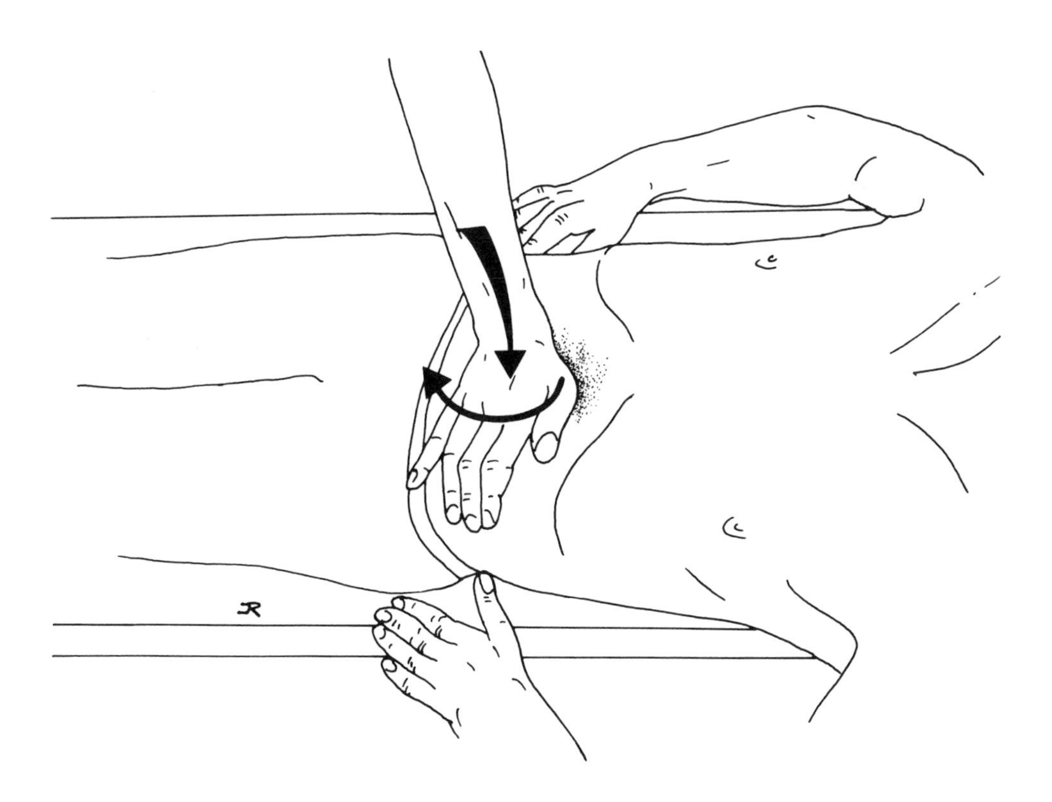

Illustration 6-9
Induction of the Sphincter of Oddi

When this, or any of the other sphincters, are so restricted that their motion is difficult to influence by rotational techniques, we use a preparatory transverse technique. With the hand pressed down over the sphincter as described above, push medially and then laterally along a straight line. After 3-8 repetitions you will feel a local release. Continue with the rotational technique. Recoil techniques are also useful for sphincters.

Duodenojejunal Flexure

This technique is also done over a point. Place the heel of your hand on the left midclavicular-umbilical line approximately 2-3 fingers' width above the umbilicus. Between the hand and the flexure are found the anterior abdominal wall, greater omentum, stomach and transverse colon (depending upon their position) and small intestine. The left kidney is located behind the jejunum and the two structures are often manipulated together. Press the heel of the hand (or fingers or thumbs — whichever is most comfortable for you and the patient) down gently, as far as possible. Ease the pressure until a rotational motion is appreciated. Follow this rotation, usually either predominantly clockwise or counterclockwise, until a still point followed by a release occurs. Finally, encourage the normal clockwise rotation of the flexure. Treitz's muscle seems to act as a sphincter, which produces a rotational motion, and also orients the flexure in order to increase and speed-up the transit of materials. We believe that it is possible to influence both functions of this muscle with induction.

Jejunoileum

Place the palms of your hands on the abdomen in the same way as for the motility test. Apply a rotational pressure following the inherent motility, with an added component directed from bottom to top for the vertical portion *(Illustration 6-10)*. Follow this until a release is appreciated. If it is difficult to perceive motility, use some mobility techniques first. With the jejunoileum, this technique usually treats upsets from bending of the lumen angles, spasms or microadhesions. It also facilitates fluid circulation of the small intestine.

Adjunctive Considerations

The treatment of an ulcer demands a regional approach. For either gastric or duodenal ulcers, the treatment should start at the esophageal junction, proceed through the stomach to the pylorus, and include the gall bladder, duodenum and sphincter of Oddi. Then treat the same structures in the reverse order. For a lasting effect, it is absolutely necessary to include the gall bladder. We primarily use motility when treating ulcers.

The right kidney often loses its motility following a duodenal ulcer and therefore needs to be treated at the same session. In cases of major restrictions, we advise you to stretch the psoas because of its close relationship with D3, D4 and the kidneys. These stretching techniques have proved very effective.

ASSOCIATED OSSEOUS RESTRICTIONS

Associated osseous restrictions are less precise or systematic than those of the liver and stomach. One finds T12 and L1 restrictions for the duodenum more often on the right than on the left. Restrictions for the jejunoileum go from T10 to L2.

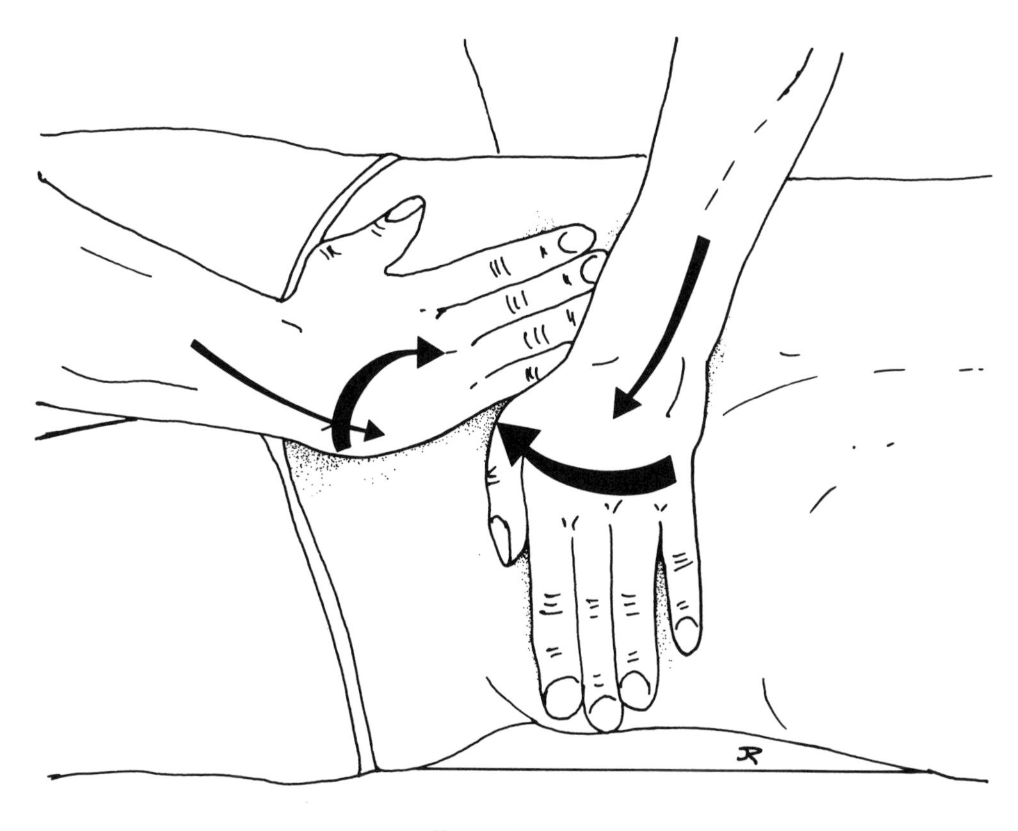

Illustration 6-10
Induction of the Jejunoileum

RECOMMENDATIONS

We ask patients to reduce their protein consumption, particularly in the evening, because experience has shown that the intestines have trouble with too much protein and do not work well at night if meat or cheeses are eaten in the evening. We also ask them to split up meals during the daytime, to apply heat locally, and to lie in the reverse Trendelenburg position when the stomach is empty. We begin by treating the liver and biliary system, then the duodenum, jejunoileum, cecum and colonic flexures.

Chapter Seven: The Colon

Table of Contents

CHAPTER SEVEN

The Colon

The colon extends from the cecum to the rectum, with an average length of 1.5m. Its diameter diminishes progressively, from 7-8cm in the ascending colon, to 5cm in the transverse colon, to 3-5cm in the descending and sigmoid colon. In the lower rectum there is an area of dilation, the rectal ampulla. The ascending and descending colons are retroperitoneal.

It is important to know how to manipulate the colon, particularly where there are relatively sharp angles: the ileocecal junction, hepatic flexure, splenic flexure and sigmoid angle. These are areas of lesser circulation where there can be risk of inflammation. The cecum is the area which most often requires manipulation because of appendectomies followed by varying degrees of healing. Restrictions here can also have an effect on the urogenital organs. Manipulation of the colon always has some effect on the kidneys. Other viscera which have a close relationship to the colon are detailed below.

Anatomy

RELATIONSHIPS

The colon comes into contact with many different structures as it travels along the sides, top and bottom of the abdominal cavity *(Illustration 7-1)*. These anatomical relationships are one reason why the colon plays a major role in visceral manipulation.

The cecum is a cul-de-sac, open at the top, which is approximately 6cm long and 5-7cm wide and can hold over 200cm^3 of material. It is located in the right iliac fossa and goes obliquely inferiorly, medially and anteriorly. On its left superior and slightly posterior aspect is the ileocecal valve, a fissure whose upper and lower lips project into the cecal cavity. Structurally, this valve is derived from the invagination of the ileum (except for the longitudinal fibers) into the cecum. Anteriorly, the cecum relates to the abdominal wall. When empty, it may be separated from the wall by loops of the small

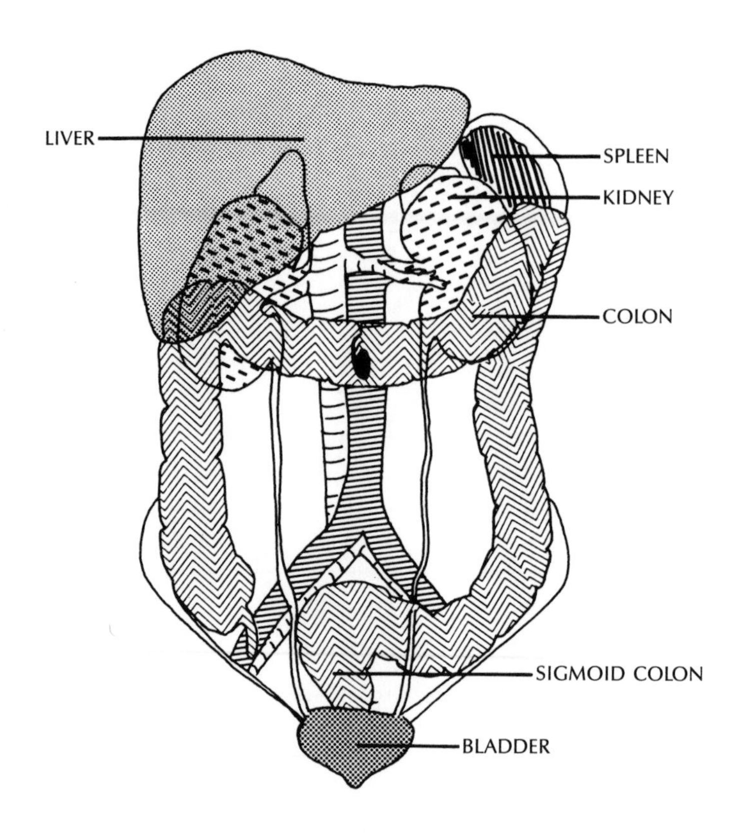

Illustration 7-1
Anatomical Relationships of the Colon

intestine. Posteriorly, it rests on the parietal peritoneum and comes into contact with other structures. These include the fatty subperitoneal layer (which helps fill the space along the inguinal ligament), iliac fascia, sheath of the external iliac vessels, inguinal ligament, psoas muscle, femoral and genitofemoral nerves and a fatty layer between the iliac fascia and the muscle containing the latter nerve. Laterally, the cecum comes in contact with the soft parts of the iliac fossa inferiorly and the lateral abdominal wall superiorly. Medially, it is in contact with the intestinal loops, particularly the end of the jejunoileum, and goes along the anteromedial side of the psoas.

The vermiform appendix is a vestigial structure, located 2-3cm above the ileocecal junction, and measures 5-10cm in length. It is usually found along the medial aspect of the cecum. Its position varies, but we believe that the descending position is the most common. The only consistent point is the opening into the cecal ampulla, situated on the anteromedial wall of the cecum, 2-3cm below the ileocecal valve. This is the deep aspect of McBurney's point. Posteriorly, the appendix rests on the iliac fascia, psoas muscle and peritoneum; with irritation, inflammation can cause psoasitis, more commonly on the right than on the left. The appendix comes in contact with the anterior abdominal wall, the small intestine posteriorly, the inferior aspect of the cecum laterally and the spermatic and external iliac vessels inferiorly. It sometimes contacts the organs of the pelvis (ovaries, rectum and bladder); a rectal examination is an essential part of the clinical examination in cases of possible appendicitis.

The ascending colon rises superiorly and slightly posteriorly so that the superior extremity is more deeply situated. It is slightly concave anteriorly, and its anterior, lateral and medial aspects are covered by peritoneum. Posteriorly, it relates to the iliac fossa, lumbar fossa and inferior end of the right kidney. Toldt's fascia separates it from the iliac fascia, aponeurosis of the quadratus lumborum and perirenal fascia. Laterally, it relates to the lateral abdominal wall and diaphragm. Medially, it relates to the right ureter, spermatic or utero-ovarian vessels, intestinal loops and inferior portion of D2. Anteriorly, it relates to the anterior abdominal wall and inferior side of the liver, upon which it leaves an impression found anterior to that of the kidney.

The hepatic flexure forms an angle of 70-80 degrees, and is oriented in a sagittal plane with the opening directed anteriorly, inferiorly and medially. This flexure is found between the right kidney posteriorly and the liver anteriorly. Most of the time, it relates to the anterior extremity of R10 and the inferior side of the liver. Medially, it relates to D2. Laterally, it is attached to the diaphragm by the right phrenicocolic ligament.

The transverse colon extends superiorly and toward the left, the left extremity being higher than the right. Its curvature is concave posteriorly. The middle part of the colon is closer to the anterior abdominal wall, with the flexures being deeply situated. The transverse colon can take on any form — tracing an "M," "S," "U," "V," "W," etc. These variations are often due to changes in the splenic flexure, which is more mobile than the hepatic flexure. The transverse colon is in contact with the liver and the abdominal wall via the greater omentum. The fixed segment is in contact superiorly with the liver and the mobile segment with the greater curvature of the stomach, up to the spleen. Posteriorly, it is over the right kidney and D2. The mobile segment is linked to the wall by the transverse mesocolon and rests on the pancreatic head, D3/D4, jejunoileum and left kidney.

The splenic flexure forms a more acute angle (roughly 50 degrees) than that of the hepatic flexure, and is oriented anteromedially on an oblique sagittal plane. Compared to the hepatic flexure, it is deeper, further away from the body's medial axis and higher (at the level of R8). Anteriorly and medially, it contacts the greater curvature of the stomach, beyond which it goes to rest upon the diaphragm. It relates superiorly to the spleen and left phrenicocolic ligament, and laterally to the diaphragm, lateral abdominal wall and ribs.

The descending colon begins at the splenic flexure and ends at the level of the iliac crest. It is smaller and further posterior than the ascending colon, and is located in a groove lateral to the left kidney. Anteriorly and laterally, it relates to the loops of the small intestine. Posteriorly, it is in contact, through Toldt's fascia (a dense connective tissue covering the colon posteriorly), with the lateral edge of the kidney and the posterior abdominal wall.

The upper part of the sigmoid colon, which is fixed, begins at the posterosuperior part of the internal iliac fossa, and goes from top to bottom following the external edge of the left psoas until 3-4cm from the inguinal ligament. There it bends and crosses the anterior face of the psoas to enter the pelvic cavity. The part of the colon between the iliac crest and the true pelvis is sometimes called the iliac colon. It is in contact posteriorly with the iliac fascia, Toldt's fascia and the external iliac vessels which go along the internal edge of the psoas; anteriorly, it contacts the jejunoileum. The middle portion of the sigmoid can be as wide as 15-16cm. The lower part of the sigmoid colon begins at the medial edge of the left psoas muscle and ends at the rectum. At the right edge of the pelvic cavity, it bends upon itself to go obliquely inferiorly, posteriorly and medially, joining the rectum at the level of S3. Inferiorly, it is in contact with the bladder and rectum. In women, it contacts the bladder, uterus and in some cases the vesicouterine or rectovaginal cul-de-sacs.

We do not manipulate the rectum during treatment of the colon, but it is important in treatment of the coccyx (see chapter 11).

VISCERAL ARTICULATIONS

The different parts of the colon contact and connect with many important structures. The most mobile parts of the colon are the transverse and sigmoid segments. They are linked to the posterior abdominal wall by connective tissue which permits some movement (the ascending and descending segments are fixed more firmly by Toldt's fascia). It is upon these mobile parts that turgor and intracavitary pressure (see chapter 1) have the greatest effect.

The cecum is mobile and attached superiorly to the posterior abdominal wall by a peritoneal fold. Inferiorly and medially, it is attached by the inferior part of the mesentery.

The ascending colon is usually held against the lumbar fossa by the peritoneum, which surrounds it on three sides and is reinforced by Toldt's fascia posteriorly. Sometimes, there is also a mesentery which provides a certain freedom of movement.

The hepatic flexure is maintained by the peritoneum, reinforced by three specialized peritoneal folds:

- the right hepatocolic ligament arises from the inferior side of the liver to insert itself in the hepatic flexure and the anterior side of the right kidney
- the cystoduodenal ligament, a prolongation of the lesser omentum, goes from the bile duct to the duodenum and hepatic flexure
- the right phrenicocolic ligament links the diaphragm to the hepatic flexure and often extends along the transverse mesocolon and greater omentum

Despite these attachments, the hepatic flexure can move. We want to emphasize the important and close relationship it has with the liver and the kidney.

The transverse mesocolon is a mesenteric fold connecting the transverse colon to the posterior abdominal wall. Very short where the flexures are, it is as long as 15cm in its middle part. It forms a horizontal partition between the stomach and small intestine. The parietal edge is oblique superiorly and from right to left. The transverse mesocolon crosses over the anteroinferior aspect of the right kidney, upper third of D2 and the pancreatic head above the duodenojejunal flexure and ends at the diaphragm with the left phrenicocolic ligament.

The greater omentum, which links the stomach to the transverse colon, is in front of the jejunoileum and just behind the anterior abdominal wall. It connects to the diaphragm laterally via the phrenicocolic ligaments.

The transverse colon is more mobile on the right than the left, despite the presence of the gastrocolic ligament (part of the greater omentum) on the right. The splenic flexure on the left is connected with the diaphragm and the lateral abdominal wall by the left phrenicocolic ligament. The latter is the principal attachment, reinforced by the descending colon.

The descending colon, like the ascending colon, is attached to the posterior abdominal wall by Toldt's fascia. Less commonly, it is also connected by a mesentery.

The sigmoid mesocolon forms a posteriorly and inferiorly concave curvature. The parietal insertion is shorter than the visceral insertion. It goes from the posterior edge of the iliac crest, then inferiorly, anteriorly and medially, crossing the psoas muscle. There is an exchange of fibers with the root of the mesentery. It follows the medial edge of the psoas, going superomedially to the level of L4/5. It then folds upon itself, runs inferomedially, crossing over the primary iliac artery, and carries itself like a median axis through the level of L5/S1 to L3 where it ends.

Sliding Surfaces

These are extremely numerous and we will not name them all. We wish to emphasize the fact that the colon is most mobile in its transverse and sigmoid portions, and that it has a very close relationship with the kidneys. Manipulations of one of these organs necessarily has an effect on all the others.

TOPOGRAPHICAL ANATOMY

The cutaneous projection of the cecum is a triangular area in the iliac fossa bordered by the midclavicular line medially, the line connecting the top of the iliac crests superiorly, and the inguinal ligament inferiorly and laterally. The most common cutaneous projection of the base of the appendix is McBurney's point, which is one third of the distance from the anterior superior iliac spine (ASIS) to the umbilicus. Also to be noted is Lanz's point, which forms the anatomical seat to the base of the appendix. It is located one third of the way from the right ASIS to the left ASIS, and is often higher in men than in women. Remember that the position of the appendix can vary widely. The cutaneous projection of the ileocecal junction is located at the intersection of the midclavicular line with the line connecting the two iliac tubercles. This is usually 2cm superior to McBurney's point.

The hepatic flexure, located deeper than the cecum, has an anterior projection at R10. The splenic flexure, situated deeper and higher than the hepatic flexure, is also further away from the median line and its anterior projection is at R8.

The reference points for the transverse and sigmoid colons vary greatly according to respiration, digestion and the fullness of the surrounding organs. For example, the pelvic loop of the sigmoid colon is often pushed out of the pelvis by a full bladder, rectum or uterus, or by its own fullness. It can therefore be found either at the pelvic brim or in the left or even the right iliac fossa. Normally, the transverse colon is found between two transverse planes, the upper one passing through the ninth costal cartilages and the lower one through the umbilicus. It can descend as far down as the pelvis.

Physiologic Motion

MOBILITY

We are not going to study all the varieties of intestinal movement during respiration. Only the colonic flexures, directly linked to the diaphragm by the phrenicocolic ligaments, will be considered here. We have used fluoroscopy to study this mobility. Fortunately, there is air in the colon which makes the flexures more visible during radiography. During inhalation, the following movements occur:

In the frontal plane, diaphragmatic movement is of greater amplitude laterally. The flexures follow the cupula and move inferiorly and slightly medially by approximately 3cm. With forced inhalation, this movement can be as much as 10cm.

In the sagittal plane, because of the diaphragmatic push, the summits of the flexures move anteroinferiorly. Overall, during inhalation, the flexures move from top to bottom, front to back and laterally to medially.

Aside from these movements of the colonic flexures, the transverse colon also moves inferiorly in a frontal plane with varying excursion depending on its state of fullness; the fuller it is, the higher its resting position.

MOTILITY

Locally, each part of the colon makes a transverse motion on its parietal connecting fascia (Toldt's fascia), which creates in turn a medial and lateral, or (for the transverse colon) a superior and inferior, frontal concavity. There is also a local rotational motion around the longitudinal axis of the colon. The resultant motion has both translatory and rotational components *(Illustration 7-2)*. It is our experience that the flexures are more affected by mobility than motility. The ileocecal junction finds itself between the movements of the small intestine and the colon, and makes a counterclockwise rotation in expir or when stuck in the closed position.

There is also a large-scale rotation in which the small intestine and colon follow the direction of formation of the digestive tract and, particularly, intestinal twisting during embryogenesis *(Illustration 7-3)*. This movement, in a clockwise and counterclockwise furling and unfurling manner, is as described for the small intestine (pages 146-47), except that its starting point for the colon is the cecum.

Indications for Visceral Evaluation

These are primarily transit problems. When colonic transit is disturbed, chyle and other materials stagnate, causing local irritative phenomena ranging from irritable bowel problems to colitis, which ultimately can lead to infection. Constipation is a result of colonic atony, with multiple causes ranging from simple negligence in diet or lifestyle to major hormonal instability. Some cases of constipation have mechanical origins — notably adhesions resulting from an appendectomy. In these cases we have often obtained excellent results.

Special attention should be paid to the angles of the colon — the cecum, the hepatic and splenic flexures and the sigmoid colon — which are zones of reduced circulation (particularly on the left). Parasites are often found in these angles. Muscles of the colon can go into spasm, presenting an obstacle to good transit and fluid circulation. A spasmodic colon, also known as irritable bowel syndrome, is a good indication for our manipulations.

In women it is not uncommon for the right ovary and the cecum to be connected by a ligamentous structure and, in some cases, to lean against each other. While congestion of the ovary can adversely affect the cecum, we believe that it is more common for

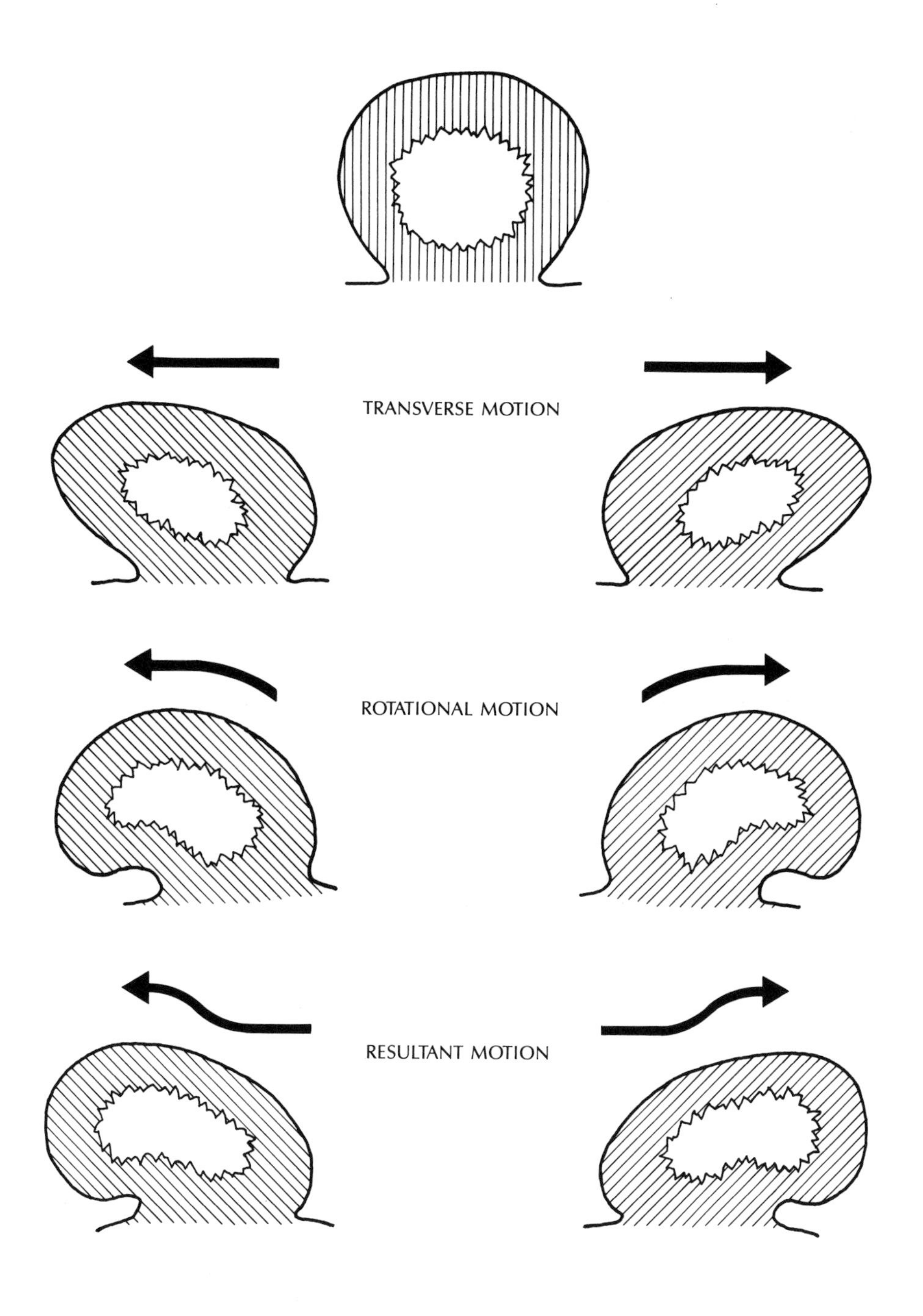

Illustration 7-2
Localized Motility of the Colon

cecal irritations to affect the ovary. This, and the effects of appendectomies, are possible reasons why ovarian problems are much more common on the right.

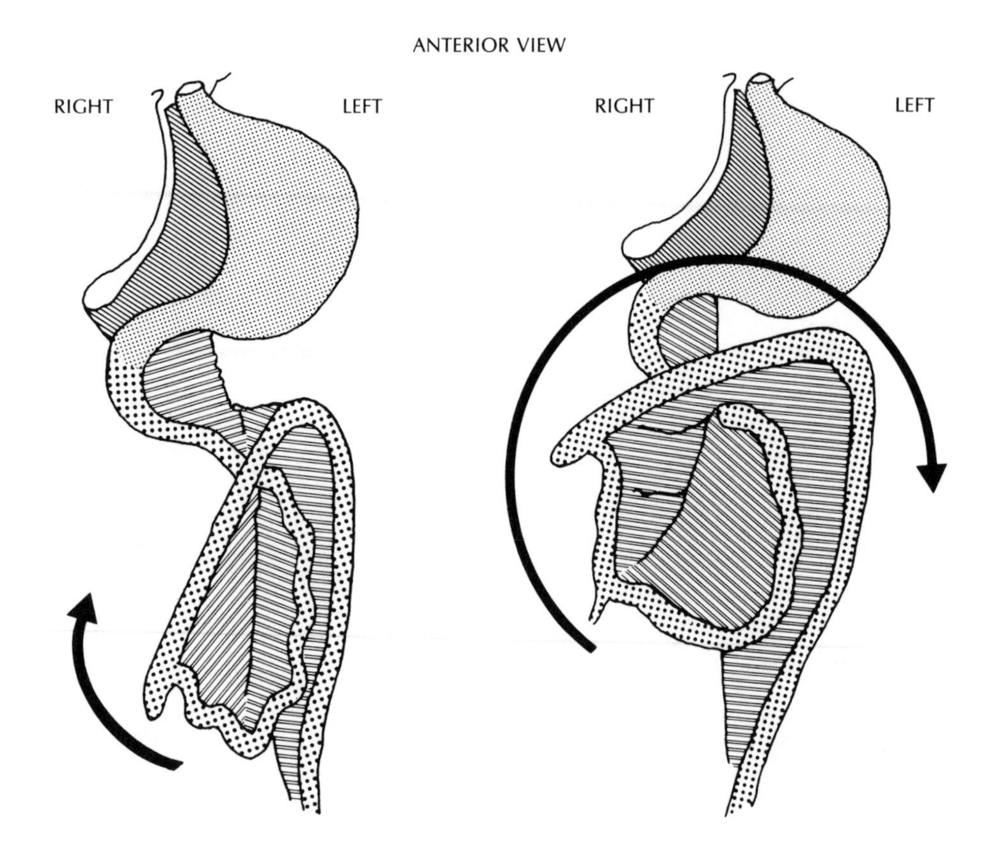

Illustration 7-3
General Motility of the Colon — Anterior View

Also, do not forget the intimate relations the colon has with the kidneys and liver, particularly the former. The connection between the cecum and right kidney is very important. We will go so far as to say that anytime you find it necessary to evaluate the right kidney you must evaluate the cecum as well.

Evaluation

Once again, please refer to your clinical manuals. Know how to gather information about the frequency, consistency and color of bowel movements. If these are black or red, more specialized examinations are called for, including radiography and endoscopy. Do not hesitate to refer to outside consultations. For example, rectocolic neoplasms are common but may be difficult to diagnose. A diagnosis of appendicitis is one of the most difficult to make, as most surgeons will confirm, because the appendix has a variable position and has been found in extraordinary places. The symptoms of appendicitis are sometimes confusing and require minute exploration — if in doubt, use the rectal exam, or send the patient to the emergency room.

Percussion and particularly palpation are important sources of information. The colon has the virtue of being palpable along almost its entire length. During the exam, you must also search for scars. External scars are often only the superficial tip of an extensive internal system of adhesions.

Besides problems with the colon itself and those of contiguous organs, restrictions of the colon should be considered as a possible etiology for more distant conditions. For example, restrictions of the cecum or sigmoid colon may affect the psoas muscle and/or obturator nerve. This can lead to hip and knee problems, in some cases mediated by the articular branch of the nerve which innervates the articular capsule. Other problems which commonly involve restriction of the colon include impotence, varicose veins and sciatica.

To decide if sciatica on the left is related to restrictions of the sigmoid colon, perform the straight-knee leg raising test. If it is positive (Lasègue's sign), put your finger over the sigmoid mesocolon and push it toward the umbilicus (this is a mild form of the treatment technique described below). If the degree of leg-raising is unchanged, a disc problem is likely; if the leg can be raised higher, the epidural veins may be involved secondarily to problems with the sigmoid colon; if the leg cannot be raised as high as before, the sciatica is more likely to be caused by a nerve irritation. A similar procedure can be performed for sciatica on the right and the cecum. The kidneys may also be involved and can be tested in a similar fashion.

PALPATION

Palpation of the cecum is carried out in the supine position with the hips and knees bent. Approach the cecum gingerly but thoroughly. After locating it in the iliac fossa, palpate both the medial and lateral aspects. Then press down slowly until you can feel the body of the cecum. When you feel it, be sure to evaluate the tension of the body of the cecum as well as its posterior, inferior and lateral attachments. The ascending colon is directly palpable via the abdomen, without the interposition of any other organ. The easiest place to start is directly to the right of the umbilicus, lateral to the duodenum. The descending colon is palpable across the loops of the small intestine and greater omentum. The sigmoid colon is accessible across the urachus and the loops of the small intestine over the bladder. These parts of the colon are also palpated best in the supine position.

To palpate the hepatic and splenic flexures, we prefer the kyphosed seated position which permits deeper penetration into the subdiaphragmatic region. The hepatic flexure is accessible by placing the fingers under the liver (colonic impression) between L2 and the right phrenicocolic ligament in front of the kidney. It is more difficult to reach and identify the splenic flexure. Situated higher and deeper, it is easily palpable only in thin patients via the greater curvature of the stomach.

MOBILITY TESTS

With the patient lying supine and with knees bent, or in the knee-elbow or lateral decubitus positions, check the mobility of the cecum medially, laterally and superiorly. The cecum should move easily in all these directions. It is important to check the posterior attachments as closely as possible, which is most easily done in the knee-elbow position.

To test the ascending and descending colons, place the patient in the supine position with legs bent. Move these segments like the string of a bow, creating medial and then lateral concavities. The colon should be elastic and return quickly to its original position. Palpation may be sensitive but should not be painful. Palpating in this manner will indicate the areas of colonic spasm, which will be difficult to "pluck." Areas of relative atony will deform easily but return slowly.

To test the hepatic and splenic flexures, place the patient in a kyphosed seated position, with your fingers positioned as for liver or stomach manipulation but as far as possible from the median line. For the hepatic flexure, use the same type of movement as that used for testing the liver (pages 92-93). For the splenic flexure, use the same movement as for the gastric fundus and upper part of the greater curvature of the stomach (pages 124-25), except that the fingers are as far to the left as possible under the rib cage. This is a difficult test for novices to perform; to make it easier, have the patient sidebend to the left and slightly rotate the thorax to the right once the fingers are well positioned. This will enable the fingers to go as high as possible.

To test the pelvic mesocolon, place the patient in the supine position, legs bent and feet resting on a cushion. Stand on the patient's right and apply a posterior pressure just medial to the left iliac crest and superior to the inguinal ligament after having appreciated the anterior abdominal wall, greater omentum and small intestine with the pads of the fingers. When the fingers are deeply situated (without pain), pull the abdominal mass toward a line inferior to, but parallel with, the root of the mesentery. You should be able to appreciate a thin, firm area which is the root of the sigmoid mesocolon. Areas of too much tension indicate a restriction or adhesion.

It is difficult to find a specific test and manipulation for the transverse colon. Its relationships are forever changing. In order to affect it, we usually manipulate the flexures. Sometimes the middle part can be affected via manipulation of the small intestine. For a mobility test of the mesenteric root, see page 153.

MOTILITY TESTS

For these tests, stand beside the supine patient and place the fingers of your left hand on the ascending colon (palm on the cecum), and the fingers of your right hand on the descending colon (palm on the sigmoid angle).

The most important of the local tests concern the ileocecal junction and sigmoid colon. During expir, the cecum makes a clockwise rotation accompanied by a movement taking its superior aspect superomedially. The sigmoid colon rolls up upon itself with a clockwise rotation and moves toward the umbilicus during expir *(Illustration 7-4).* The ileocecal junction should have a cyclical motion. If it is stuck in one motion or the other, there may be a general spasm, or the sphincter may be stuck open or in inspir (clockwise motion), or closed or in expir (counterclockwise motion).

The general motility of the large intestine is the same as that of the small intestine; one cannot dissociate the two. During expir, the entire intestinal tract makes a large-scale clockwise rotation and the cecum and the sigmoid colon move superomedially. This is a motion of great amplitude and is relatively easy to find.

Restrictions

The most common restrictions are due to operative sequelae and inflammations. The ileocecal junction, which is normally mobile, is often fixed after appendectomies;

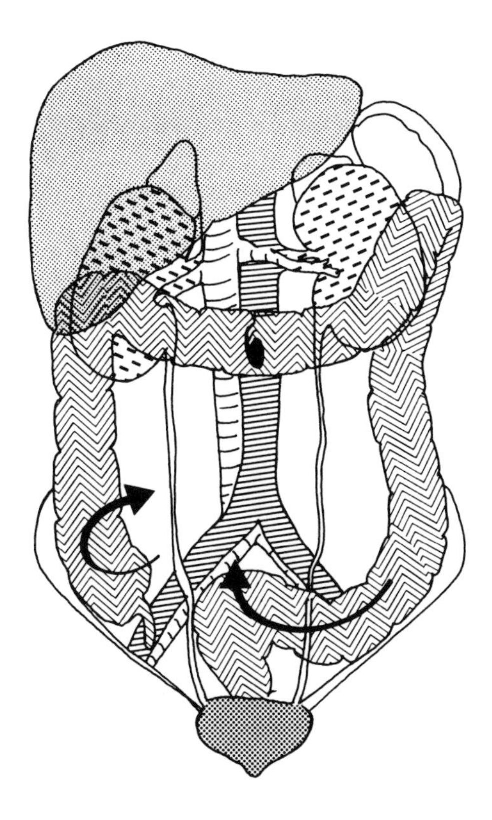

Illustration 7-4
Ileocecal and Sigmoid Motility — Expir

there will be a rotation and subsequent chronic spasm and decrease in function. A laparotomy can have the same restrictive effect on the colon. Problems due to surgeries require direct techniques, as induction will not work.

If the cecum is fixed, it may pull the right kidney, right ovary or peritoneum to the right, with many associated adverse effects. This is particularly a problem during pregnancy. Slight peritoneal inflammations, which are very common and do not necessitate surgery, have a negative effect on intestinal mobility and motility by provoking microadhesions. Restrictions of the colon may also lead to sciatica.

A restriction of the cecum can lead to a stretched or spasmed ascending colon. Besides affecting the right kidney, there is also an effect on the liver. Usually this causes the anterior liver to be drawn inferiorly, with a resultant decrease in liver motility.

Manipulations

DIRECT TECHNIQUES

It is best to begin each treatment session with the ileocecal region which, when stimulated, seems to awaken the entire colon via the mesenteric plexus.

Cecum

Place your thumbs either on the lateral right third of the line connecting the two ASIS's, or on the inferior third of the line between the umbilicus and the right ASIS, depending on the position of the cecum (which should be palpable). The cecum often appears to be lower in women than men because of differences in the shape of the pelvis. In turn push the lateral side of the cecum superomedially, the medial side inferolaterally and the inferior side superolaterally *(Illustration 7-5)*. This technique can be done in

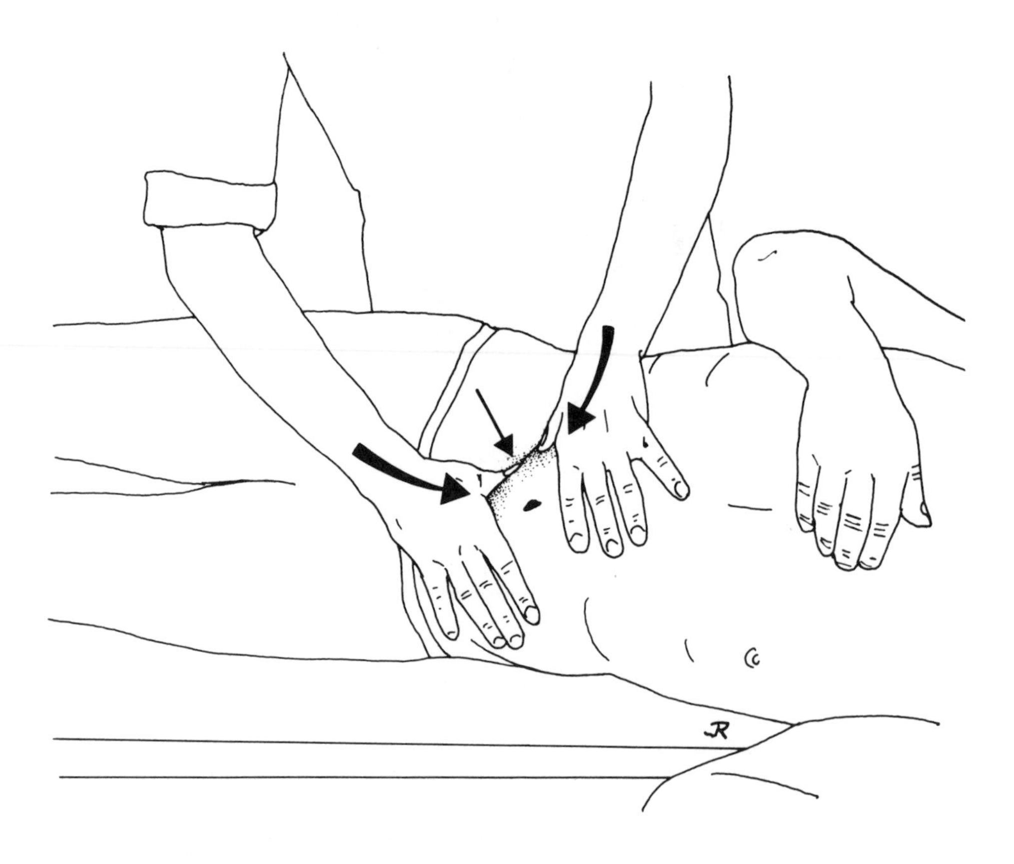

Illustration 7-5
Direct Manipulation of the Cecum — Supine Position

either the supine or right lateral decubitus position. With the latter, the fingers can more easily go deep into the abdomen, which makes this technique particularly effective *(Illustration 7-6)*.

For a stronger effect, push your hands gradually toward the back in order to get at the posterior aspect of the cecum. If this can be mobilized, the effect is quite dramatic. Another way to more easily probe deep into the abdomen is to have the patient assume the elbow/knee position which relaxes the abdominal wall. This position is useful for direct techniques on any part of the colon.

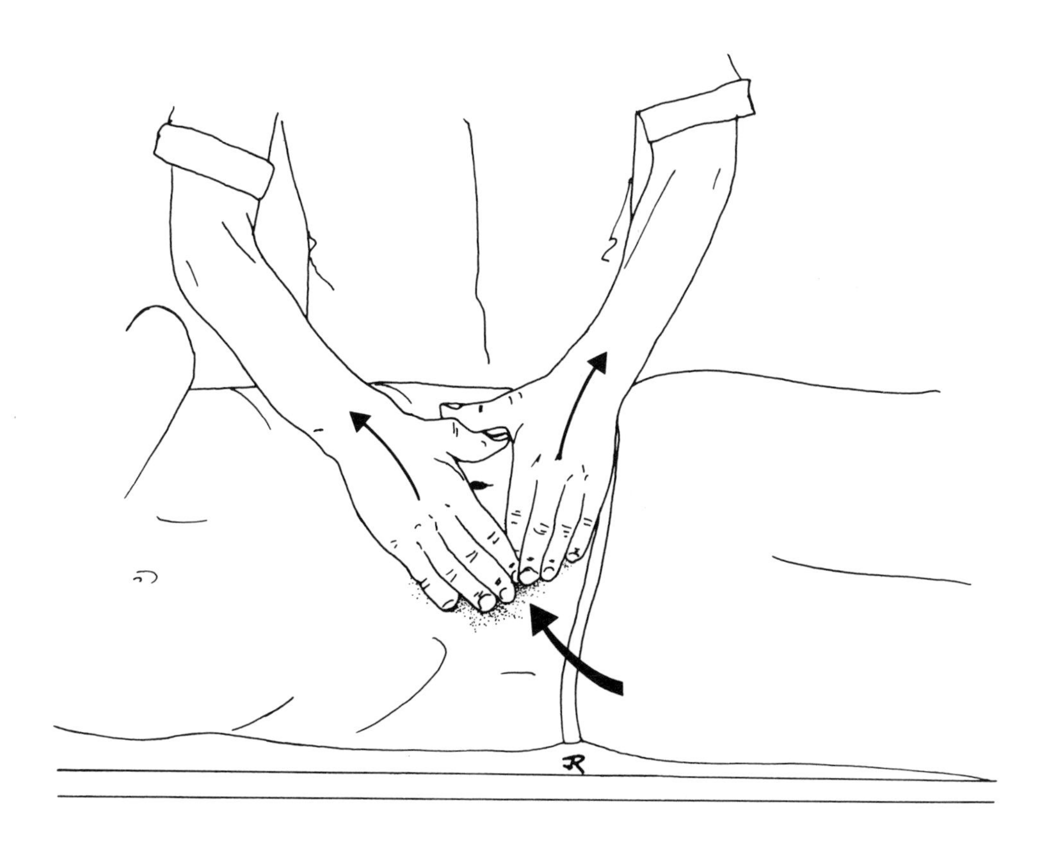

Illustration 7-6
Direct Manipulation of the Cecum —
Right Lateral Decubitus Position

It is necessary to release all cecal restrictions as well as the ascending colon before releasing the ileocecal junction. To do this, place the thenar or hypothenar aspect of the heel of the hand over the surface projection of the ileocecal junction, which is usually 2cm above McBurney's point. Then perform the direct pressure sphincter-release technique (page 153), followed by induction of the sphincter (page 155).

Ascending and Descending Colons

With the patient in the supine or lateral decubitus position (left lateral for the ascending and right lateral for the descending colon), insert your fingers between the lateral abdominal wall and the colon in order to alternately push it toward the umbilicus and let it come back, in a gentle rhythmic manner *(Illustration 7-7)*. It will help to simultaneously spread your thumbs apart as you push. If you are able to go deep posteriorly and work on the attachments to the mesocolon, the treatment will be much more effective. Concentrate your efforts on areas that are restricted.

For the lower and posterior parts of the colon, have the patient lie in the lateral decubitus position. Then push the colon medially by placing your thumb on the area

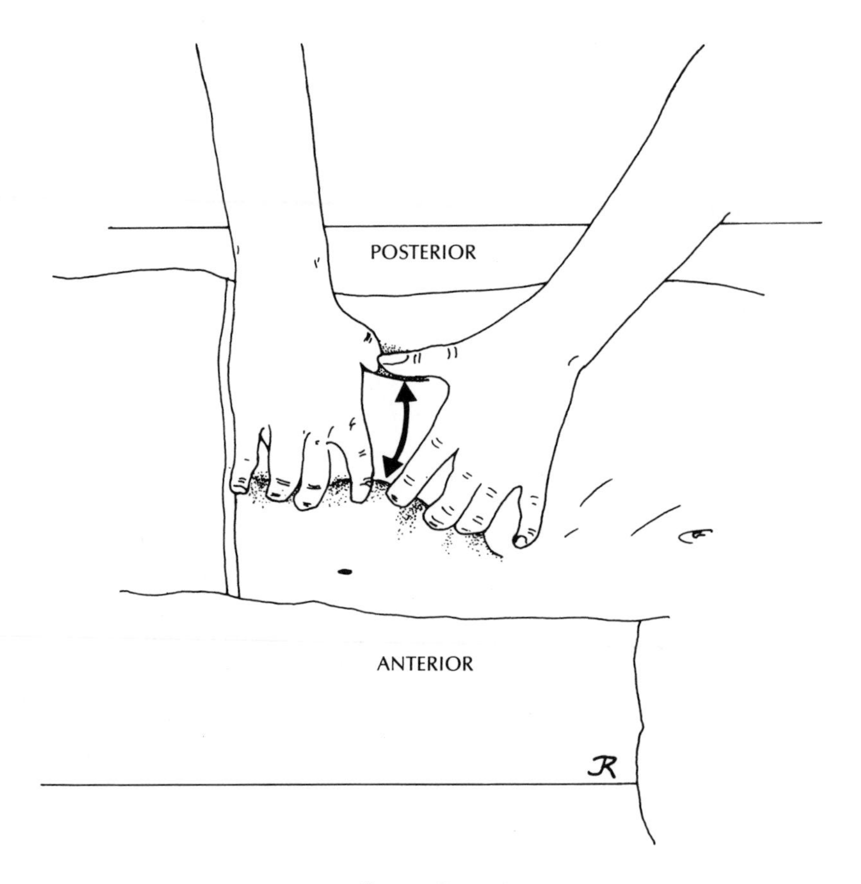

Illustration 7-7
Direct Manipulation of the Ascending Colon —
Left Lateral Decubitus Position

between R12 and the iliac crest. This is the only movement which enables you to rather easily release restrictions of the posterior part of the colon.

Hepatic Flexure

The patient should sit in a marked kyphosed position because the hepatic flexure is posterior and this position makes it easier to penetrate deeply. Place your fingers between the area where it becomes difficult to feel the ascending colon and the rib cage (usually facing R10), and push them posteriorly and laterally under the hepatic region. With practice, you can feel directly the hepatic flexure and the right phrenicocolic ligament. The technique consists of stretching the ascending colon by drawing this flexure upward and slightly medially *(Illustration 7-8)*. This also affects the transverse colon, for which there is no specific technique. The hepatic flexure can also be treated indirectly by lifting the liver. This is because the hepatocolic ligament which connects the two structures enables you to stretch the ligament via the liver.

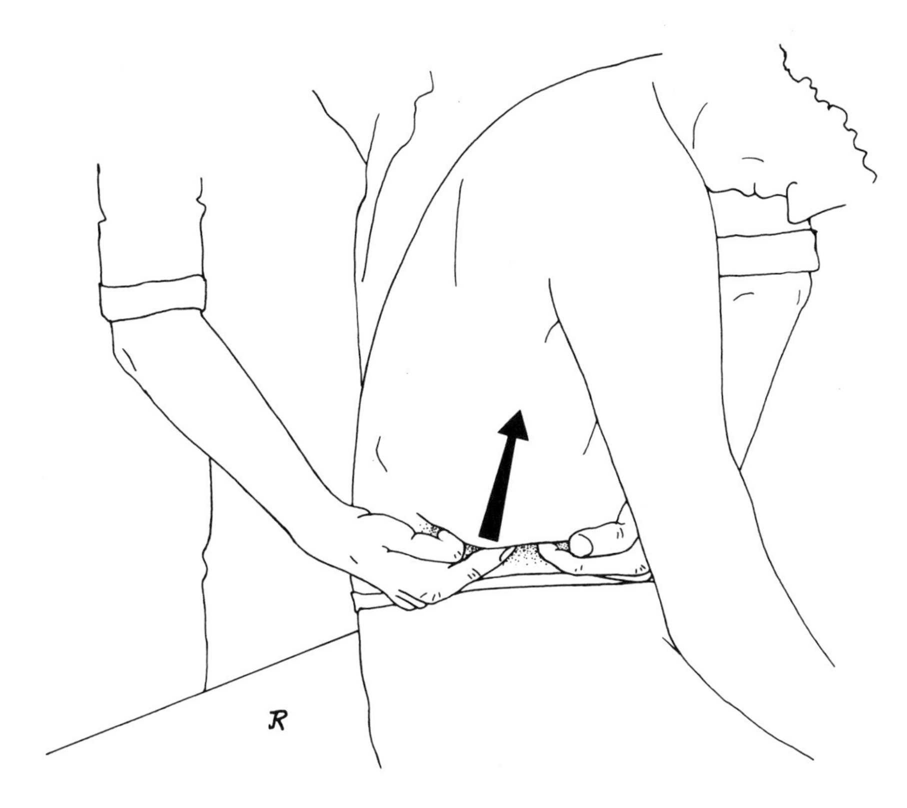

Illustration 7-8
Direct Manipulation of the Hepatic Flexure of the Colon — Seated Position

Splenic Flexure

With the patient sitting in a marked kyphosed position, place your fingers facing R8. The splenic flexure is very mobile and therefore difficult to reach. It is attached to the stomach by a part of the left phrenicocolic ligament. In order to manipulate this flexure, a technique similar to that for the gastric fundus (page 129) is performed, with the fingers directed as far to the left as possible. The splenic flexure is often full of air which may be confused with the stomach's air pocket. (The wall of the colon is much thinner and softer than that of the stomach.) The technique consists of drawing the flexure upward and laterally to stretch the descending colon, stomach and transverse colon via the gastrocolic ligament.

Sigmoid Colon

This technique is directed at the sigmoid colon and the left psoas muscle. With the patient in the supine position, legs bent, place your fingers on the left lateral psoas, 3-4cm from the inguinal ligament, and draw the small intestine, sigmoid colon and its mesocolon superomedially in the direction of the umbilicus. It is necessary to push your fingers in first and then direct them superomedially. Repeat this technique medial to the left psoas *(Illustration 7-9).* Sometimes placing the patient in a left lateral decubitus

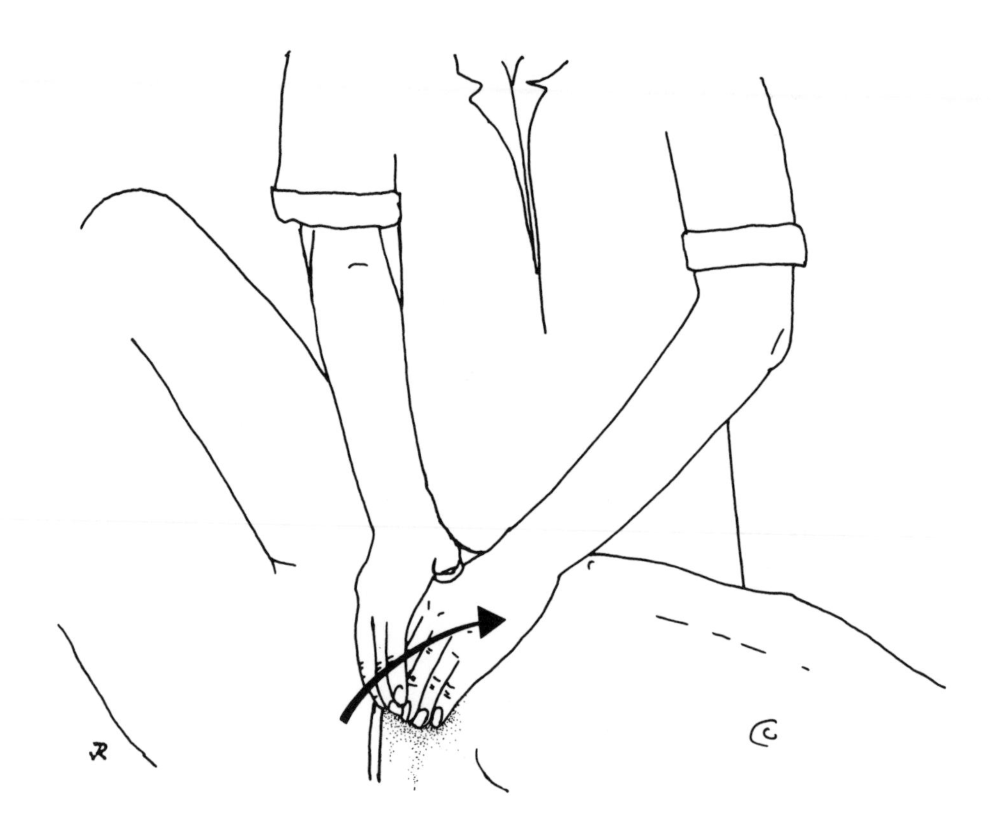

Illustration 7-9
Direct Manipulation of the Sigmoid Colon — Supine Position

position will improve results *(Illustration 7-10).* Extension of the left hip joint will stretch the psoas to varying degrees. The part of the colon which passes anterior to the psoas is sensitive and, if it is involved, must be explored and manipulated. The sliding surfaces are often affected in inflammatory phenomena. Direct pressure is used; techniques should be slow, gentle, progressive and painless. Fecal matter, which is hard in the sigmoid colon, can be easily felt. Treatment of the sigmoid colon should be preceded by treatment of its mesocolon. This is merely the addition of stretching (through the hands) and local mobilization (through the fingers) over areas of restriction found with the mobility test (page 172).

The pelvic portions of the small intestine and colon often rest upon the uterus and bladder, even in the cul-de-sacs if those organs are full and push the sigmoid colon upward. For manipulation of this segment, the patient must be in the supine position, legs bent and feet on a cushion. Place your fingers above the point on the pubic symphysis

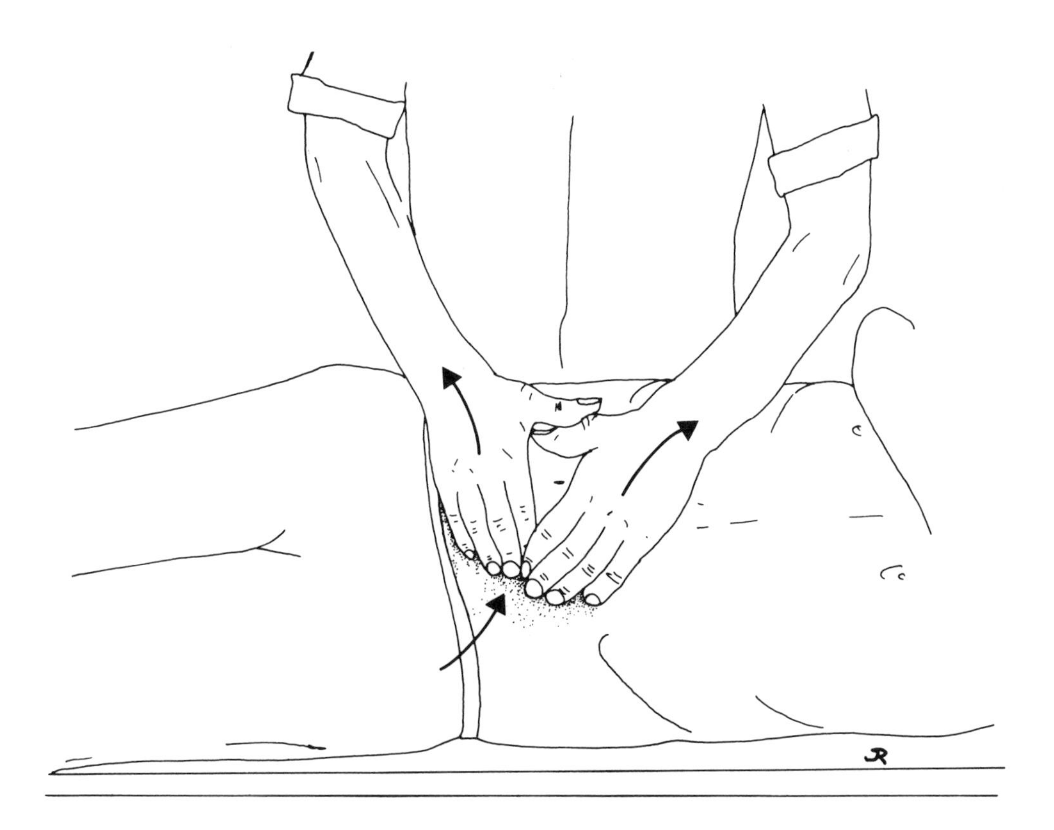

Illustration 7-10
Direct Manipulation of the Sigmoid Colon —
Left Lateral Decubitus Position

corresponding to the upper part of the bladder and drive the small intestine and the colon superiorly, toward the umbilicus. This technique is also used for the bladder, as the weight of the colon can contribute to prolapse of the bladder.

COMBINED TECHNIQUES

The combined techniques consist of manipulating the colon with the assistance of movements of the trunk and head (for the flexures) or the lower limbs. With the patient in a supine or lateral decubitus position, use one hand to manipulate the colon and the other to mobilize the bent legs on the body. For example, for manipulation of the cecum, the patient assumes the left lateral decubitus position. First flex the hip so that it is easy to penetrate deeply into the abdomen with your fingers. Then use one hand to draw the cecum in the direction of the lesion (the way it goes more easily), while the other extends and externally rotates the right thigh. As the leg is returned to the neutral position, increase the pressure on the cecum. Rhythmically continue this procedure until you perceive a release.

In a similar manner, you can manipulate the ascending, descending and sigmoid colons. For example, draw the sigmoid colon upward and toward the right with one hand, while the other hand pushes the legs downward and toward the left *(Illustration 7-11).* The efficacy of this technique is enhanced by performing it in the reverse Trendelenburg position. These are very effective techniques because the opposing movements of the legs and your hand increase the stretching effect. We often use this technique for the lower abdominal and pelvic organs.

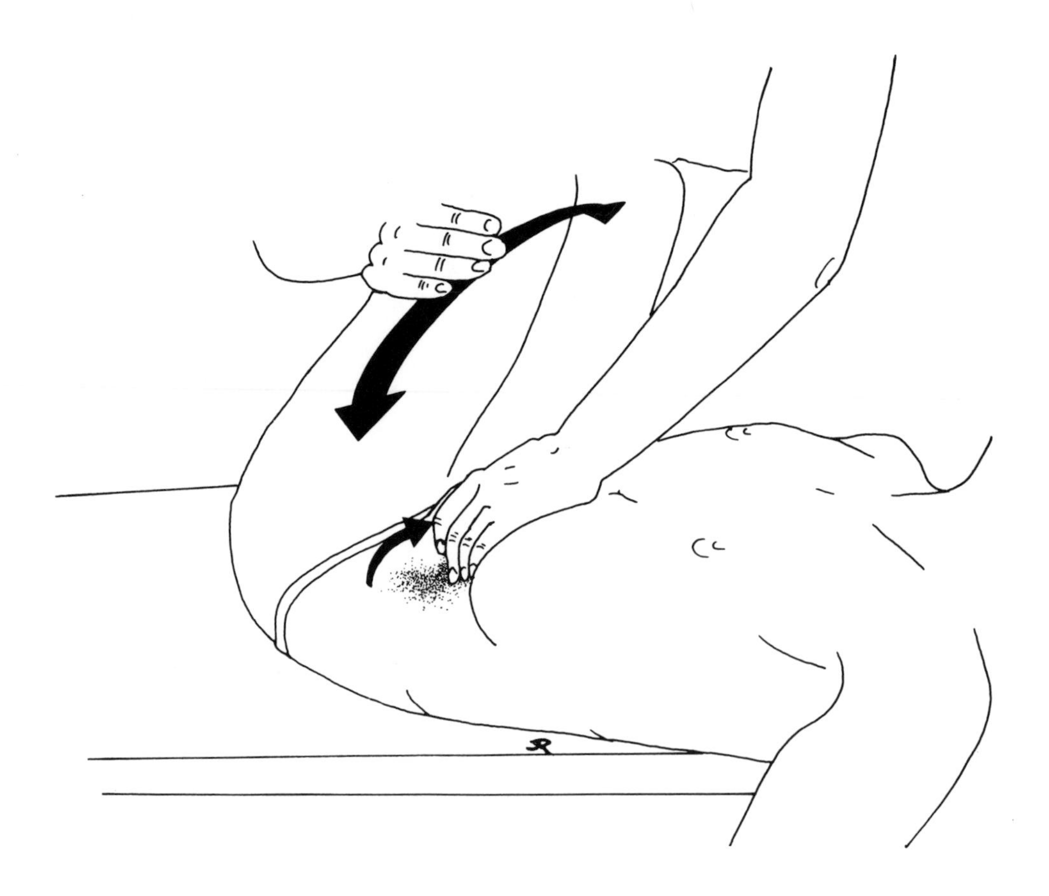

Illustration 7-11
Combined Manipulation of the Sigmoid Colon —
Supine Position

In order to manipulate the flexures with the patient in a seated position, it is helpful to adjust the position of the thorax by combining flexion and sidebending movements in order to obtain the optimal localization of forces. Consider the example of the splenic flexure. Once your fingers are placed as far as possible to the left in the hypochondriac region, increase flexion in order to allow the fingers to go farther toward the back; next, left sidebend and right rotate the patient so that your fingers can go as far superiorly as possible *(Illustration 7-12).* Repeat this rhythmically until there is a release; then repeat

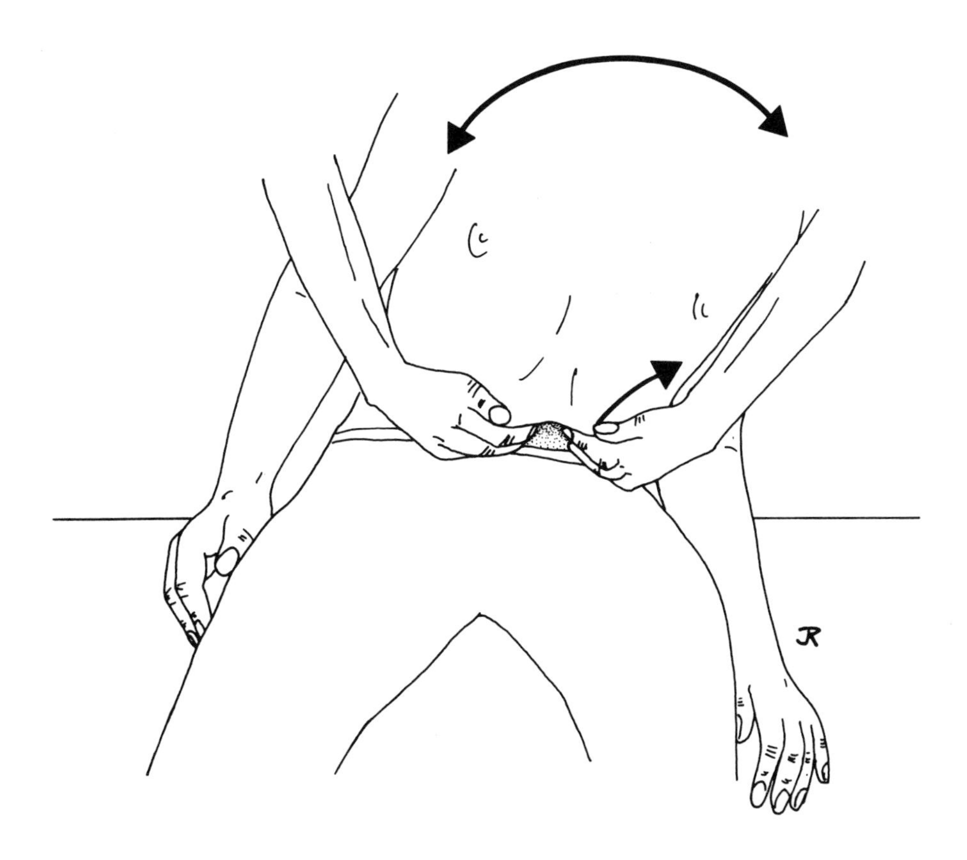

Illustration 7-12
Combined Manipulation of the Splenic Fixure of the Colon — Seated Position

with more flexion and posteriorly-directed pressure. Continue until the restriction disappears. The possibilities for fine-tuning your forces in this way are limited only by your imagination and palpatory skills. With this technique, we have seen movements of the colon of up to 6cm on fluoroscopy.

A variation of this type of technique can provide a general stretch to the ascending or descending colon. In the kyphosed seated position, fix the flexure you are working on with one hand and the cecum or inferior end of the descending colon with the other. Then extend, sidebend and rotate the trunk in such a way as to put maximum stretch on the area being treated. Relax and repeat rhythmically until you feel a release with both hands. This is an excellent technique to use to finish up a treatment of colonic mobility.

The following technique, which acts on both flexures simultaneously, is a good concluding technique. With the patient in the kyphosed seated position, put one hand under each costal border with the fingers pushed posteriorly on the flexures. As you extend the patient, pull back on your hands; as you flex here, push the ribs in. Repeat this rhythmically and gently until you appreciate a release. This technique also has an effect on the transverse colon.

INDUCTION TECHNIQUES

For general induction, as in the motility test, you stand beside the supine patient and place the fingers of your left hand on the ascending colon (palm on the cecum), and the fingers of your right hand on the descending colon (palm on the sigmoid angle). During expir, both hands simultaneously perform a clockwise movement in which the left hand moves superomedially while the right moves inferomedially *(Illustration 7-13)*. Reverse these movements during inspir. In some cases, the motion will be stuck in one phase when there is a problem. The treatment consists of following and encouraging the predominant part of the cycle while simply following the other part with a slow and gentle rhythm until there is a release (with or without a still point) and resultant smooth motion. All treatments using localized techniques should be finished by general induction. Induction also affects the small intestine which, as previously noted, is embryologically formed with the colon. The movement should be done slowly and fully.

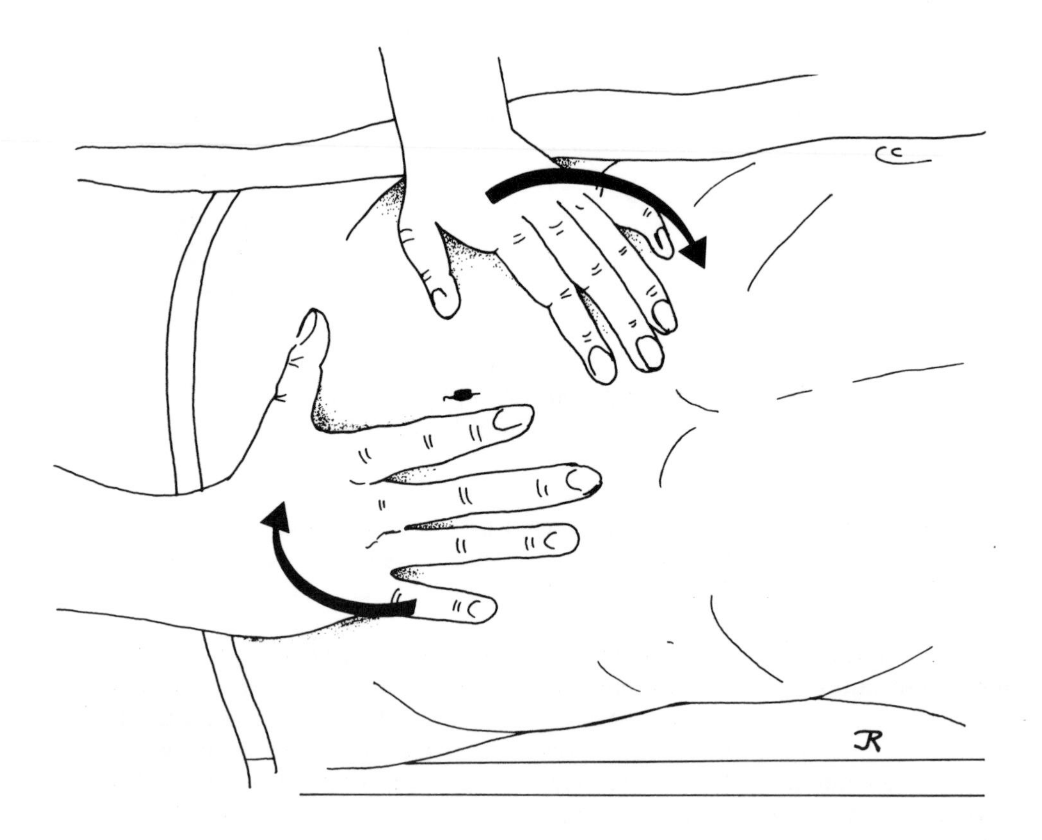

Illustration 7-13
General Induction of the Colon — Expir

Local techniques are employed for the ileocecal region and for localized spasms in any area; the purpose is to reduce spasms of the colonic muscles. Apply rotational pressures with the palm or heel of the hand on the affected region; follow the clockwise or counterclockwise motion (whichever predominates) and encourage it until a release is felt. This local induction technique is the same as that described above for the pylorus, duodenojejunal flexure or sphincter of Oddi (page 155).

Effects

We have drawn several conclusions from the application of these techniques:

- manipulation of the ascending colon always affects the small intestine, right kidney and liver
- manipulation of the descending colon always affects the stomach, left kidney and small intestine
- manipulation of the sigmoid colon always affects the small intestine, bladder and genital organs

In cases of right oophoritis without apparent cause, pay particular attention to the cecum — there is a close connection between the appendix and the ovary which may be at the root of the problem.

Adjunctive Considerations

ASSOCIATED OSSEOUS RESTRICTIONS

With problems of the colon, one often finds associated lower lumbar and sacroiliac restrictions. These are not definite indications for manipulation if they are secondary to visceral problems. The simple fact of freeing colonic tensions is often enough to improve or dissipate these restrictions. You should manipulate the viscera first and then, only afterward, the vertebral column.

RECOMMENDATIONS

The patient must learn to eat lightly in the evening, when the colon has more difficulty in the digestion and transit of foodstuffs. Fat, meat and sugars are most likely to cause problems. During treatment, it is better that the patient ingest food with long fibers (e.g., leeks, spinach, celery, swiss chard, etc.) in order to improve the general function of the colon. Like the small intestine, the large intestine is dependent on the liver and pancreas, and it is a good idea to stimulate these organs (with olive oil, lemon or herbs) during the course of treatment.

Chapter Eight: The Kidneys

Table of Contents

CHAPTER EIGHT

The Kidneys

The kidneys are a very important, almost key organ in visceral manipulation. Early anatomists thought that they were fixed and incapable of movement, and that all ectopic kidneys and prolapses were of congenital origin. We now know that this is far from the truth. In fact, renal ptoses are relatively common.

During our studies and early years of practice, we heard many patients with cystitis, hypertension or other symptoms report having had a kidney "lifted" by some old bonesetter and that this technique had in some way eliminated the problem. We smiled indulgently the first few times we heard these stories, but as the examples multiplied we became intrigued by them. Positive results in such cases may be viewed by overly conventional health practitioners as examples of the "placebo" effect. We preferred to investigate and understand such cases, rather than dismiss them out of hand.

Osteopaths in Europe are a fringe group in medicine, but the border between the mainstream and the fringe keeps changing (both as the mainstream expands and the fringe moves farther out), and we therefore decided to try out these techniques. Having used them successfully, we now wish to transmit them to a wider audience, and at the same time to explain their effectiveness. We have learned that kidneys move naturally and can be manipulated. A fixed kidney is pathological.

When we began our studies, scans were not available, but intravenous pyelograms (IVP) were sufficient for our purposes. We chose patients with an ectopic kidney. Typically, during the IVP, we could see on the control screen that the ectopic kidney was not only placed oddly but also moved less than the other kidney. After manipulation, fluoroscopy showed that while the ectopic kidney remained roughly in the same position, it now moved in the same manner as the other kidney, i.e., was in good functioning condition. This and other experiments have led us to the belief that mobility is of greater significance than position. We would like to again thank Drs. Arnaud and Roulet of Grenoble, without whom we would never have been able to construct this theory.

According to tradition, a kidney can only be palpated by a posterior route. Again, our experience contradicts this established idea. In fact, there are two acceptable routes for palpating a kidney. Palpation of a kidney from the back takes place across a musculoskeletal wall 8cm in thickness. Palpation by the anterior route takes place across 1.5cm of muscles and approximately 10cm of mobile viscera. It is obvious to us, based on this anatomy and our clinical experience, that the anterior route works better for both the practitioner and the patient.

Anatomy

It is of vital importance that you understand the relationships of both kidneys in order to know where to place your fingers. The kidneys are deeply placed in the lumbar region (or kidney region) of the abdomen, on each side of the T12/L1 junction. They are retroperitoneal and surrounded by perirenal fat. They appear to be nestled into nests of adipose tissue, and held up by blood vessels which go in and out of their hili (medial aspects). The average kidney is about 12cm long, 7cm wide and 3cm deep and weighs 140g. The left kidney is often larger (especially longer), while the right one is placed a little (about 1.25cm) lower down. We shall see why when we study their relationships.

Appropriately, each kidney has the shape of a kidney bean. The longitudinal axis is slightly oblique from top to bottom and medially to laterally. The anterior and posterior sides, respectively, face slightly laterally and medially. The concave edge faces medially and slightly anteriorly and the convex edge faces in the opposite direction.

RENAL FASCIA

The subperitoneal tissue which presses the kidneys against the posterior abdominal wall thickens to create a fibrous sheath called the renal fascia. At the level of T12/L1 the fasciae of the two kidneys fuse together in front of the vertebrae. Ptosis of one kidney often affects the other via this connection. The renal fascia is divided into anterior and posterior lamina which meet at the top and sides of each kidney.

The posterior lamina covers the quadratus lumborum and psoas muscles and fixes itself to the anterolateral vertebral column medial to the psoas. It is strong, of a pearly color and attaches itself to the diaphragm. According to Gerota (a 19th-century anatomist who performed many dissections), it separates the quadratus lumborum from its aponeurosis by means of an adipose pararenal layer.

The anterior lamina follows the parietal peritoneum which it lines. It covers the anterior side of the kidney, the hilum and the large prevertebral vessels. It is thinner than the posterior lamina but is reinforced in the places where it is in contact with the colon by an area of fibrous connective tissue called Toldt's layer. This reinforced area is larger on the left than on the right.

The two laminae fuse above the adrenal gland. This fused sheath surrounds the adrenal capsule and adheres strongly to the inferior side of the diaphragm.

At the level of the inferior pole of the kidney, the two laminae approach each other but do not merge. They are lost in the adipose tissue of the internal iliac fossa. Some modern anatomists claim that in fact the two laminae fuse inferiorly and medially, entirely enclosing the kidney. Our impression is that, although variation does exist, the renal fascia is generally open at the bottom, forming a funnel-shape in which the kidney can be engulfed.

PARARENAL BODY AND RENAL MEMBRANE

The kidney has no peritoneal tunic. Behind the renal fascia, there is a considerable quantity of fatty tissue called the pararenal body, which is the adipose capsule of the kidney. It first appears around ten years of age. It is semifluid in consistency and located mostly on the lateral aspect of the kidney, as well as the inferior and posterior aspects. It "melts" during weight loss, which increases the risk of ptosis.

There is also a fibrous membrane which adheres to the outside of the pararenal body. This renal membrane, which is fairly strong despite its delicacy, expands to line the walls of the hilum. It provides an envelope, comparable with Glisson's capsule of the liver, for the vessels which branch out in the kidney. The renal membrane is composed primarily of connective tissue, plus a small amount of elastic tissue.

RELATIONSHIPS

Posterior

Superiorly, the posterior side of each kidney rests on the diaphragm and its arched ligament, an extension of the psoas muscle and fibrous tissue which is stretched between R11/12. This aspect of the kidney corresponds to the pleural costodiaphragmatic sinus which extends down to the upper edge of L1.

Inferiorly, the posterior kidney rests on soft tissue between R12 and the iliac crest (the left kidney also comes in contact with R11). It contacts the psoas and its iliac fascia laterally, and the quadratus lumborum and its aponeurosis medially. The latter is separated from the posterior lamina of the renal fascia by the pararenal body, which is traversed by the twelfth intercostal, iliohypogastric and ilioinguinal nerves. This explains why renal problems are often accompanied by radiating pain to the groin, hip bursae and superomedial thigh.

Muscles of the abdominal wall, especially the transversus abdominis, interact with the posterolateral kidney. The lumbar and Grynfeltt's triangles (weak points of the iliocostal wall) are found in this area.

Anterior

The right kidney is the one most often fixed or prolapsed. Anteriorly, the right kidney is connected by the hepatorenal ligament to the inferior aspect of the liver, upon which it leaves an impression. Farther down, the kidney relates to the hepatic flexure of the colon. The short mesentery crosses here and adheres to the anterior lamina of the renal fascia. The mesentery also adheres to the left kidney, but much higher up. Medially, D2 is separated from the right kidney by fascia. These arrangements mean that the right kidney is mostly submesocolic. There is a strong connection across the peritoneum between the right kidney and the ascending colon.

The anterior left kidney contacts the adrenal gland superomedially and the spleen (on which it leaves an impression) superolaterally. Between these areas, the left kidney relates to the stomach, from which it is separated by the peritoneum of the omental bursa. The middle part of the anterior left kidney relates to the pancreas. The inferior part relates primarily to the duodenojejunal function, the jejunum and, laterally, to the splenic flexure of the colon. The attachment here is much stronger than that of the hepatic flexure to the right kidney.

VISCERAL ARTICULATIONS

There is, in fact, no ligament or mesentery which has the function of keeping the kidneys in place. Furthermore, the hilar vessels and ureters are the only structures offering continuity with the kidney. The hilar vessels, being transversely stretched, can only be considered as a brake. Cruveilhier wrote in 1852: "The kidneys are surrounded by a kind of atmosphere and seem suspended by vessels."

How are the kidneys kept in place? There is no renal cavity, analogous to the thoracic or abdominal cavities, and therefore no suction system. However, as discussed in chapter 3, the viscera below the diaphragm are drawn upward by thoracic inhalation. The liver and kidneys are lighter in the living human body than on the dissection table, because thoracic inhalation tends to counteract the force of gravity. When the abdomen is opened up (as in surgery), the kidneys move relatively freely.

In the supine position, there are no forces drawing the kidney downward. In the standing and seated positions, the muscular abdominal walls are normally contracted, producing an increase in intraabdominal pressure which forces the kidneys posteriorly against the musculoskeletal wall. This tends to counteract the force of gravity, which pulls the abdominal viscera anteroinferiorly. The combination of thoracic inhalation and abdominal muscular contraction permits the kidneys to maintain an equilibrium; a person can normally permit himself to jump in the air with both feet and land without running the risk of a renal ptosis!

Sliding Surfaces

The factors favoring the kidneys' mobility also favor its pathological sliding downward. The kidney is not enclosed in a serous system with a virtual cavity. The presence of the semifluid pararenal body is another sliding factor. If the hilar vessels ran obliquely, they could function as a suspensory mechanism. Since they run transversely, this is not the case. The sliding surfaces of the kidney were described in detail on the preceding page.

TOPOGRAPHICAL ANATOMY

Because the anatomical variation of the kidneys is considerable, it is difficult to define precise reference marks. We shall therefore describe the limits instead.

Posteriorly, the superior limit is a horizontal line passing through T11. The renal pelvis is at the level of L1. The inferior limit is a horizontal line passing through L3 *(Illustration 8-1)*; but the right kidney is often found below this line.

Anteriorly, the superior limit is a horizontal line which goes through the R9 cartilages bilaterally. This line crosses the median axis at the solar plexus. The inferior poles of the kidneys are found lateral to the umbilicus (the left 1-2cm higher); they are further from the median axis than are the superior poles. In the seated position, the inferior pole on the right is often one finger's width below the level of the umbilicus. As renal ptosis is quite common, the level in different people can vary.

Some anatomists place the renal pelvises on a transverse plane which passes through the umbilicus. We think this is a mistake based on dissection of cadavers. In a living subject they are at the level of T12/L1.

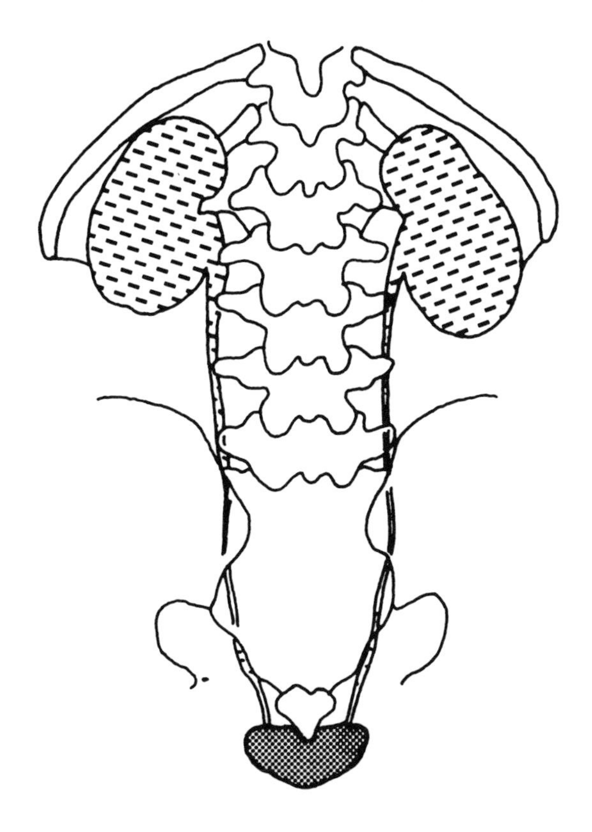

Illustration 8-1
Posterior Topographical Reference Marks for the Kidneys

Physiologic Motion

Mobility and motility occur in almost exactly the same directions and on the same axes. The direction of motion is controlled by several factors which give the kidney a certain degree of freedom:

- the renal fascia is open inferiorly and medially
- the hilar vessels diverge perpendicularly from the aorta and inferior vena cava
- the posteromedial kidney is in contact with the psoas muscle, which acts as a "rail" along which the kidney can slide down to the level of L3. Starting as a sagittally compressed structure, the psoas becomes rounder inferiorly. As the muscular slips from the lower lumbar vertebrae join it, the body of the psoas flattens out on a frontal plane.

MOBILITY

Movement is created by the diaphragm and its respiratory rhythm. During inhalation, the kidneys are pushed downward, following the "rail" of the psoas. The superior pole is pushed forward (forward bends), and the kidney moves medially to laterally (externally rotates) *(Illustration 8-2)*. The opposite movements occur during exhalation. The adrenal glands move in the same way as the kidneys.

Illustration 8-2
Mobility of the Right Kidney — Inhalation

The amplitude of this motion is roughly the height of a vertebral body (3-4cm). The motion is repeated over 20,000 times per day, for a total distance of over 600m. With certain subjects, one finds the same motion with the same amplitude when changing from the supine to the standing position. This phenomenon is easily observable with IVPs.

MOTILITY

The movements of motility are somewhat similar to those of mobility, except that we have never felt the forward rocking of the superior pole during motility. To train yourself to detect this motion, it is easier if it is split into three components. The first is the vertical rising and falling of the kidney. The second is a mild external and internal rotation. The third is a pendular motion in a frontal plane of the inferior pole around an axis which passes through the superior pole, like the movement of a windshield wiper fixed to a superior point. During inspir, the kidney goes downward and externally rotates away from the median axis *(Illustration 8-3)*. The opposite happens during expir.

Indications for Visceral Evaluation

The clinical picture of a renal pathology, or its sequelae, often develops alongside symptoms from other, sometimes far-removed, tissues. Conversely, distant pathologies can affect the kidneys. For example, renal pathology can cause problems or symptoms

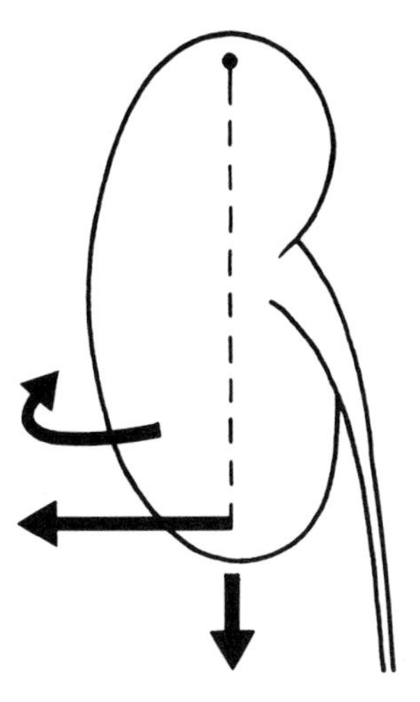

Illustration 8-3
Motility of the Right Kidney — Inspir

of the ureters, bladder, stomach or intestines, or lead to hypertension.

The most characteristic indications are the infectious syndromes such as recurring urinary tract infections. Your discriminating diagnosis should eliminate uretocystic reflux, ptosis of the bladder, infections of the genital area, etc. Infections linked with a renal ptosis have symptoms which increase with prolonged standing or sitting as the peritoneal viscera weigh down on the kidney and push it inferiorly instead of posteriorly. These cases, which are often associated with constipation, increase the downward pressure of the kidney. Symptoms will be further aggravated if two precipitants occur at the same time — a long journey in a car or train with no lavatory is a good example.

Postpartum renal ptosis is common. Because of the phenomena of suction from below and pressure from above, the kidney can easily be ptosed in the birthing process. After childbirth, the tissues are distended and the functions of intracavitary pressure and turgor are inhibited. Add to this the sudden weight loss and hypotonia which follow, and you have all the favorable conditions for a ptosis. It is a good idea to check mothers 4-6 months after childbirth.

While it is true that ptosis of the left kidney is less common than that of the right, it does occur. Its characteristic effects (e.g., varicoceles, decreased libido, impotence) usually relate to the fact that the testicular vein empties into the left renal vein in men (the right empties into the inferior vena cava). In women the left ovarian vein empties into the left renal vein; associated problems include varicosities of the left leg, labiovulvar conditions on the left and left-sided dysmenorrhea.

The list of other possible indications is very long and includes disorders of all the viscera that come in contact with the kidneys (colon, liver, stomach, etc.), as well as low back pain, sciatica and other musculoskeletal disorders of the lower extremities. For one method to test for kidney involvement in these problems, see page 171.

In our years of clinical practice, as we have gone deeper with our clinical examination and farther back in the patient's history, the path has led countless times to the kidneys. Often there is no logical anatomical or physiological explanation for the involvement of the kidneys. Osteopathy is still an empirical science; we can only hope that the mechanisms for what we observe will one day be clearly understood.

Evaluation

When the patient's symptoms give you cause to suspect a pathology of renal origin, check for the following problems: polyuria; sensations of abdominal distention accompanied by dyspnea; subdiaphragmatic or pelvic pain; a history of infections; colic from a stone; and discomfort from wearing belts or tight clothes. This list is not inclusive. Questions should be focused on prior contributory problems such as infections or obstacles to the free circulation of urine.

As always, do not hesitate to supplement your clinical evaluation with appropriate laboratory tests, including urinalysis, radiography, or scans. Manipulation of the kidney is contraindicated in cases of renal abscesses or other acute infectious pathologies.

PALPATION

While palpation of the kidneys is important, it may be impossible depending on the morphology of the patient. We have mentioned our preference for palpation via the anterior route. When the patient is in the supine position, you should place yourself by the patient's shoulder on the side opposite the kidney to be examined.

For the right kidney, we begin with a palpation of the cecum, a superficial organ occupying the lateral third of the space between the median axis and the right ASIS. With your hand in a hook-like shape, press the pads of the four fingers vertically between the loops of the jejunoileum and the superior part of the cecum (or, alternatively, between the duodenum and ascending colon). It may be difficult to push the intestinal loops away because contact is usually made at the level of the mesenteric root. Be gentle, since the iliac artery is just inside the pressure point. The hooked hand moves up along the psoas muscle, which is easily felt in longiline thin individuals. Opposite the umbilicus, you will feel a hard soap-like mass — this is the right kidney. If you are unsure, let up your pressure a bit and repeat the procedure — if it was the kidney you were feeling, it will feel the same; if what you felt was intestines compacted by your pressure, it will have disappeared. Normally, only the anterior aspect is felt, but if the kidney is ptosed the inferior pole is also palpable. This is because the ptosed kidney follows the psoas, which gets thicker as it descends, causing the kidney to become more superficial. To check mobility, push the kidney superiorly along its longitudinal axis. There should be a bit of play. With your hand in the same position, evaluate the movement of the kidney in response to breathing.

For the left kidney, start by palpating the sigmoid colon; it is covered by loops of the jejunoileum, which should be pushed medially. The colon occupies the lateral quarter of the space between the median axis and the left ASIS. Again, with your hand in a hook shape, push the pads of the four fingers past the colon, relatively far away from the left iliac artery. Move your hand upward, driving back the loops of the jejunoileum as far as possible along the psoas. At one finger's width above a transverse plane passing through the umbilicus, you can normally feel as a solid mass the inferior pole of the left kidney. As an alternative procedure, first find the duodenojejunal flexure (page 143) and go slightly medially (or laterally) to find the kidney. The left kidney is more difficult to feel than the right because it is relatively superior, and therefore relatively posterior (higher up on the psoas).

During palpation, keep away from the median line so as not to irritate the abdominal aorta; this could be dangerous for the patient. There are a few subjects with whom palpation of the kidneys is easy, in which case you should palpate it across the small intestine and greater omentum, where there is the least interference.

It is not possible to percuss the kidney directly, but you can percuss R11/12 at the back, which may cause some pain. If this is not because of the ribs, it can be a sign of a renal pathology. However, this is not a specific test because other organs, such as the liver, colic flexures and spleen, can cause a positive response to this percussion.

MOBILITY AND MOTILITY TESTS

These tests are performed in the supine position. They complement each other and are not double work. The mobility test is used mostly for determining the location of the kidney. Remember, a prolapsed kidney can be mobile but still have lost its motility. An immobile kidney has lost its vitality and its function is impaired.

For the mobility test, locate the inferior pole or anteroinferior aspect of the kidney as described above for palpation *(Illustration 8-4)*; once the pads of your fingers have contacted it, the patient breathes in and you should feel the kidney pushing inferiorly against your fingers.

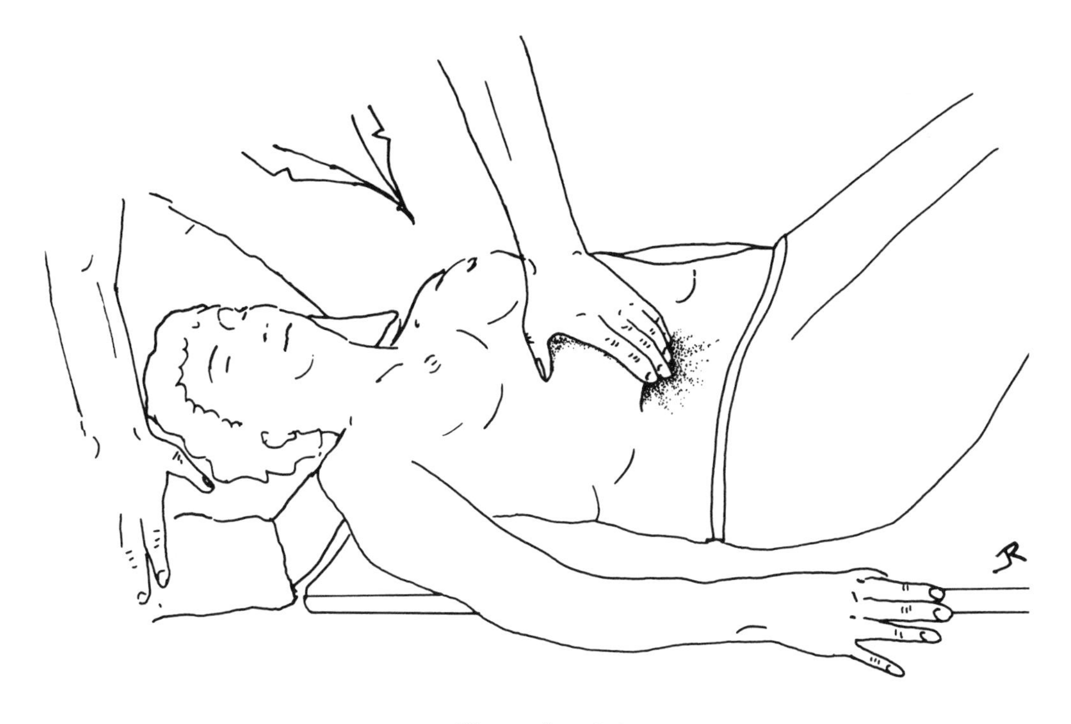

Illustration 8-4
Mobility Test of the Kidney

For the motility test, hand position is slightly different than that for mobility *(Illustration 8-5)*. Place the heel of your hand (thenar or hypothenar side, whichever is more sensitive) medial to the sigmoid or cecum (depending on the side), and push the intestine superomedially toward the umbilicus. Once the heel of your hand feels the inferior pole of the kidney (which involves an extension of your sense of touch), release the pressure on the small intestine. In order to feel the motility of the kidney, it is important that the listening hand not compress it.

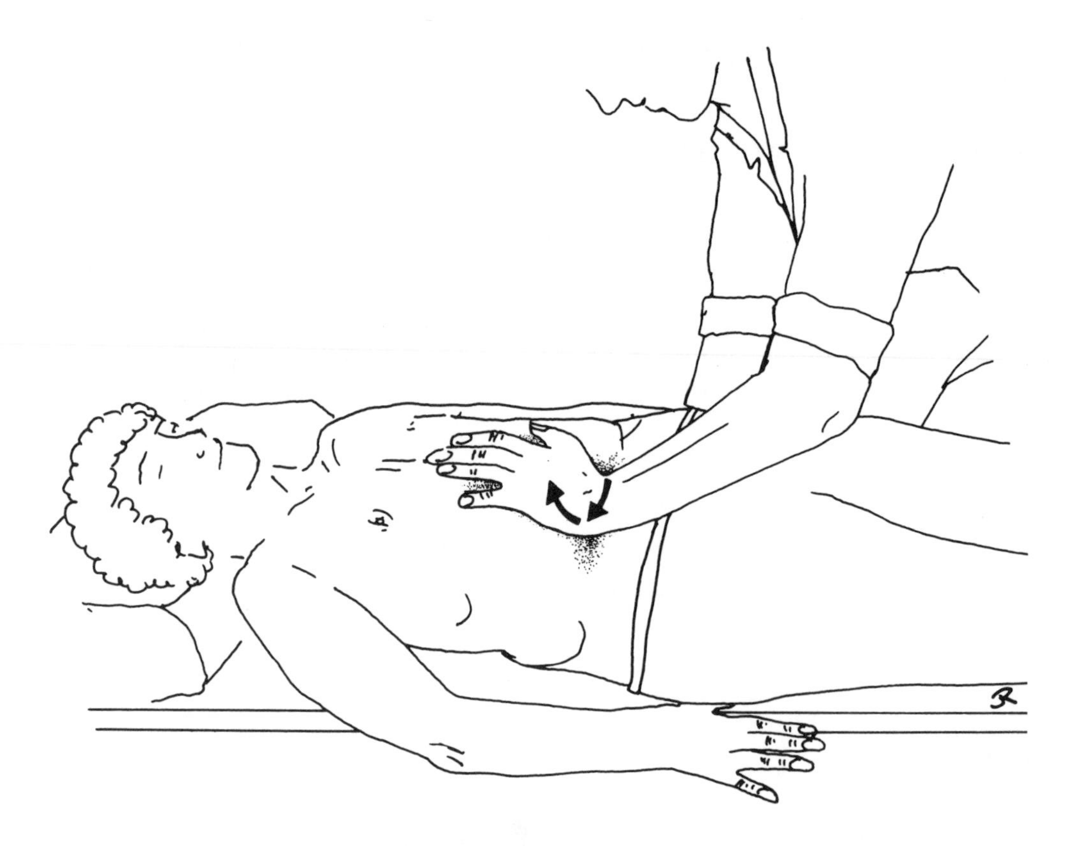

Illustration 8-5
Motility Test of the Kidney

Restrictions

Renal restrictions can be primary or secondary in nature. Judging from our experience, ptosis is the most common problem. With primary restrictions, symptoms have their cause in the kidney itself. The pathology may be congenital or acquired. Most restrictions are prolapses, the rest consisting of adhesions, postinfectious scars or the effects of surgical interventions.

Secondary renal restrictions can be of musculoskeletal or visceral origin, or be reflexive in nature. Those of musculoskeletal origin include articular restrictions of the costovertebral joints of T11/12, articular restrictions of T10-L1 or psoasitis due to muscular or bony imbalances of the lower limbs. Those of visceral origin are often due to stomach ptosis, hepatic immotility, colitis or irritable bowels. It is not uncommon to have renal motility problems that are neurological reflexes from problems with the renal centers of T6/7 and L1/2.

PTOSES

It is impossible to distinguish between congenital and acquired ptoses by clinical examination. Only a radiographic examination of the hilar vessels can make the distinction — and even then it may be uncertain. This is of no great importance since the position of the organ is less important than its functional motion or motility. Immotility arises when the "descended" kidney finds itself in a position where its motion is inhibited by pressures caused by the gravitational force on the surrounding viscera, restrictions in the adjacent tissues and/or by the tension of the abdominal muscles that maintain these viscera.

Renal ptosis is a common ailment, whether congenital or acquired. The kidneys can go as far down as the internal iliac fossa, but with these extreme ptoses the origin is congenital. It is difficult to isolate the causes of an acquired ptosis. However, they are often found in longiline asthenic individuals, and may be precipitated by trauma, violent fits of coughing, labor, marked weight loss or nervous depression. Another contributing factor is that the turgor effect which holds the kidneys in place loses its efficiency with age.

In studying the different forms of an acquired or seemingly acquired ptosis, one can see the pathway of the kidney's descent. For example, in cases of sudden weight loss, the pararenal body melts and the kidney "floats," with an increase in mobility. It follows the lateral edge of the psoas muscle. Progressively, as the psoas receives the small fibers from the transverse lumbar apophysis, it flattens out. In this way the kidney, at least the inferior pole, loses its guide. If the kidney continues downward, the inferior pole will become more medial and move anterior to the psoas. In the early stages of ptosis, the kidney will externally rotate around its pelvis. However, in more severe cases an internal rotation will occur as the kidney is held back by the tension of the hilar vessels and the shape of the psoas. The ureter, being elastic, lets itself be stretched — so much so that it may bend and twist upon itself and lose its contractility. Recurrent cystitis is another common result of renal ptosis. Remarkably, the adrenal capsule stays in place and only the kidney descends during ptosis.

Right renal ptoses are more frequent because the liver is a much more homogenous mass than the pancreas, spleen and stomach, and therefore puts more pressure on the kidney. Other reasons are that Toldt's fascia (which inserts in the colon) is smaller on the right, and that the slight left lumbar scoliosis (found in eighty percent of all individuals) brings the right kidney anteriorly, which increases the pressure of the liver.

Depending on the degree of ptosis, different nerves will be affected. With minimal inferiorly-directed tension, twelfth intercostal neuralgia may occur, with a nagging pain around R12 toward the umbilical area. With true ptosis, the iliohypogastric and ilioinguinal nerves may be involved, leading to pain from the lateral flanks toward the genital region. With severe ptosis, the genitofemoral nerve or even the lateral cutaneous femoral nerve may be irritated, producing pain which radiates from the iliac crest toward the medial knee joint.

ADHESIONS

Kidneys can be mobile and in the proper position, but with reduced motility. We use the term "adhesion" for this situation. As you listen to the motility, you do not feel the ascending and descending motion, but only a frontal rotation of the kidney around the adhesion. In patients with colitis, the hepatic and splenic flexures are often responsible for adhesions affecting the kidney. The mechanism in this case is that inflammation of the colon, via Toldt's fascia, fixes the anterior lamina and pararenal body of the kidney. Although colon-kidney restrictions are the most common, the stomach and the liver can also cause renal adhesions and fixations.

Manipulations

According to our experience, the kidney is implicated in a wide variety of visceral diagnoses. Manipulation of the kidneys will be helpful in syndromes at a distance caused by a ptosis or sequelae of renal pathologies, as well as in peripheral ailments which by mechanical, neurological or vascular connections have a repercussion on the kidneys.

In general, for right kidney dysfunction, you should treat the cecum, ascending colon, liver and hepatic flexure before working on the kidney itself. For the left kidney, release the stomach, splenic flexure and descending colon first.

DIRECT TECHNIQUES

These can be done in the supine or seated position. The procedure for locating the kidney is the same as in palpation.

Supine Position

The patient lies flat on the table with the hips and knees flexed in order to relax the abdominal wall. This position enables you to probe fairly deeply while the patient can relax completely. When treating the right kidney, we usually stand on the patient's left. Palpate the cecum with the pads of the last three fingers and then slide them medially, pushing back the loops of the jejunoileum. Pause when the fingers are one third of the way along a horizontal line joining the right ASIS to the median line, and on the lateral edge of the psoas. Shape the fingers like a hook and move them along a line that ends at the xiphoid process. At the level of the umbilicus you will usually encounter a solid mass: the anterior aspect of the inferior pole of the kidney. If there is a ptosis, you will be able to feel the inferior pole itself slightly inferior to this level. The direct technique consists of asking the patient to breathe slowly, pushing the inferior pole upward along the longitudinal axis of the kidney during exhalation, and maintaining this position as much as possible during the following inhalation *(Illustration 8-6)*. With the next exhalation, push the pole upward again. Repeat this 5-10 times until you perceive a release. At the end of the technique, release the contact at the beginning of an exhalation.

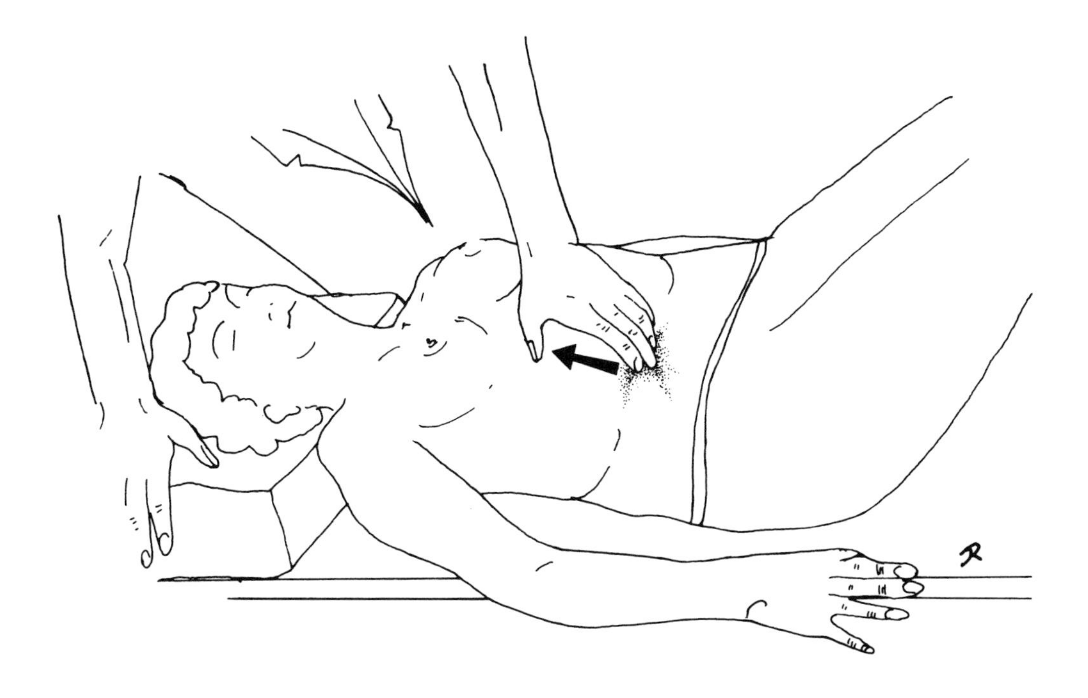

Illustration 8-6
Manipulation of the Right Kidney — Supine Position

Another way of performing this technique is to stand on the same side as the kidney being worked on. Use the heel of your hand (thenar or hypothenar side, whichever is more sensitive) and press down between the ascending colon and duodenum at the level of the kidney. The rest of the technique is the same as that described above. Remember that your upward push is directed along the longitudinal axis of the kidney. However the technique is done, be careful not to press too hard on the right iliac artery, as this can result in a vasovagal reflex or, rarely, irritate an aneurysm. This artery is found at the beginning of the exploration just medial to the contact point.

To treat the left kidney, stand at the patient's right and, with your hand in a hook shape, palpate the sigmoid colon facing the left ASIS. Push the hand upward, trying to push back as much as possible the loops of the jejunoileum. You will usually find the left kidney at one finger's width above the level of the umbilicus. Perform direct mobilization in an identical manner to that of the right kidney, contact still being made at the level of the inferior pole. For the left kidney, further precautions have to be taken for the left iliac artery. Be sure to stay approximately four fingers' width away from the median line to avoid any provocation of the abdominal aorta (you can easily feel its pulsation).

Seated Position

The relaxed patient sits astride the table. Standing behind the patient, wrap your arms around the trunk and, with the pads of the outer three fingers of each hand, make contact with the inferior pole of the kidney to be treated *(Illustration 8-7)*. In order to

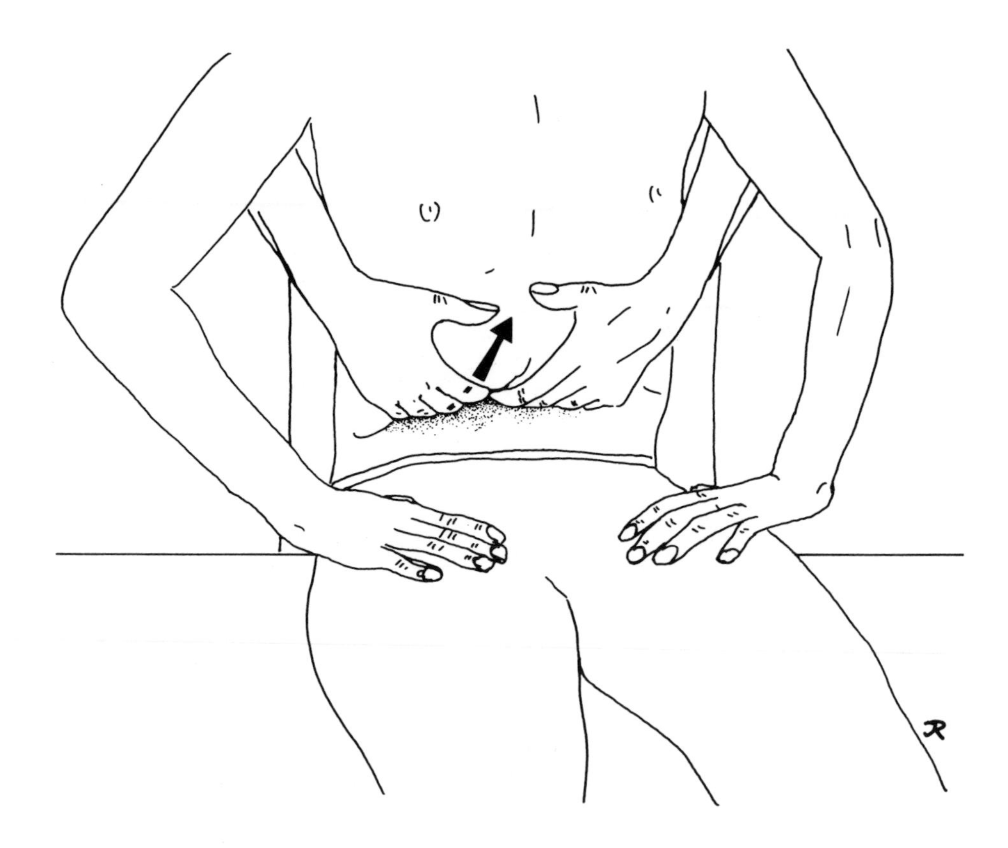

Illustration 8-7
Direct Manipulation of the Right Kidney — Seated Position

facilitate this movement, when you contact the abdomen medial to the cecum or the sigmoid colon, flex the lumbosacral junction as the patient leans backward on the ischial tuberosities. Remember to support the patient. Gradually, as the contact with the skin moves upward, localize the apex of the lumbar kyphosis higher and higher. When you are in position to mobilize the kidney, the apex should be localized at T12/L1. This enables you to have a good contact point against the inferior pole of the kidney and maximal abdominal relaxation. You can then carry out the direct technique, with exhalation, as described above for the supine position. Observe the same precautions concerning the iliac arteries and abdominal aorta.

INDIRECT TECHNIQUES

By this we mean all techniques on other organs which indirectly act on the kidney. We have seen that certain renal restrictions are caused by irritation of the colic flexures, stomach ptosis or gastritis, or inflammation of the liver, gall bladder, or duodenum. Therefore, it may be suitable to manipulate these organs as if, through them, you wished to apply suction to the kidney. When treating the right kidney, it is necessary to perform a lifting technique on the liver during the same treatment session *(Illustration 8-8)*. The

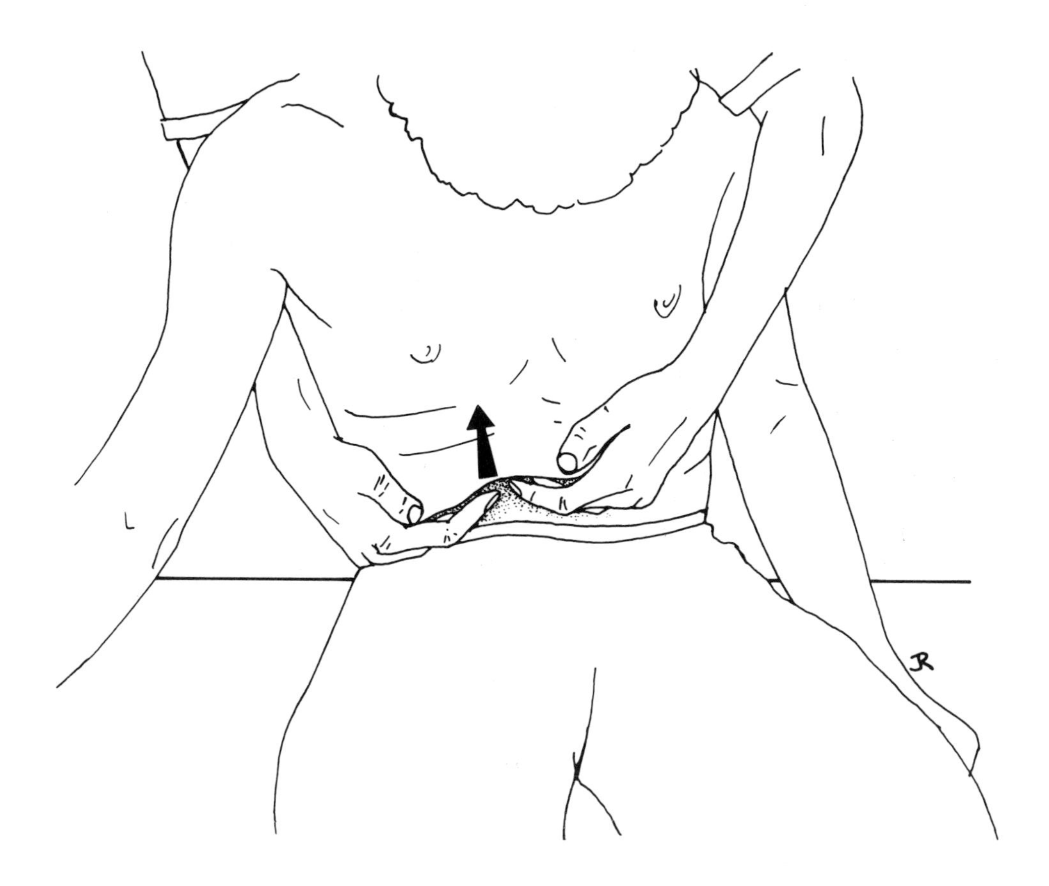

Illustration 8-8
Indirect Manipulation of the Right Kidney via the Liver

lifting movement of the stomach must always be associated with treatment of the left kidney. If the kidneys are just barely interlocking with the diaphragm, the suction effect loses its efficiency quickly. Manipulations of these other organs are discussed in the relevant chapters.

One very indirect way to mobilize the kidneys is to use direct pressure on the inferomedial surface of the navicular bone. This corresponds to the classical acupuncture point K-2 *(ran gu)*, a very important point on the Kidney channel. Although it is difficult to understand the mechanism for this from a structural viewpoint, we have often observed the release of kidney restrictions by repeated pressing of this point.

COMBINED TECHNIQUES

These are direct techniques aided by mobilization via long lever arm movements of distant structures (e.g., stretching of the psoas or quadratus lumborum), which can have an effect upon the kidney. The aim of these techniques is the direct manipulation of a kidney which is also mobilized or positioned by the movement of one or both lower limbs, the pelvis, lumbar column and/or trunk. This increases the efficacy of the technique. We shall describe three techniques as examples, but there are numerous other possibilities.

Seated Position

This is a variation of the direct technique in the seated position. To facilitate the direct pressure or manipulation of the kidney, use your arms and chest to mobilize the trunk of the patient. For example, to manipulate the right kidney, perform a rotational movement to the left which will move the right kidney forward and increase the efficacy of the treatment for renal ptosis. With the patient totally kyphosed, ask him to make successive rotational movements to the left (counterclockwise when viewed from above). With each movement, carefully stay with the inferior pole of the right kidney and move it superiorly. Gradually, as your contact point ascends, decrease the kyphosis by applying force through your thorax, bringing the trunk of the patient to the erect position at the end of the technique *(Illustration 8-9)*. This powerful movement should be done gently and should not traumatize the kidney. During the rotational movement to the left, ask the patient to breathe moderately in order to increase your contact with the kidney.

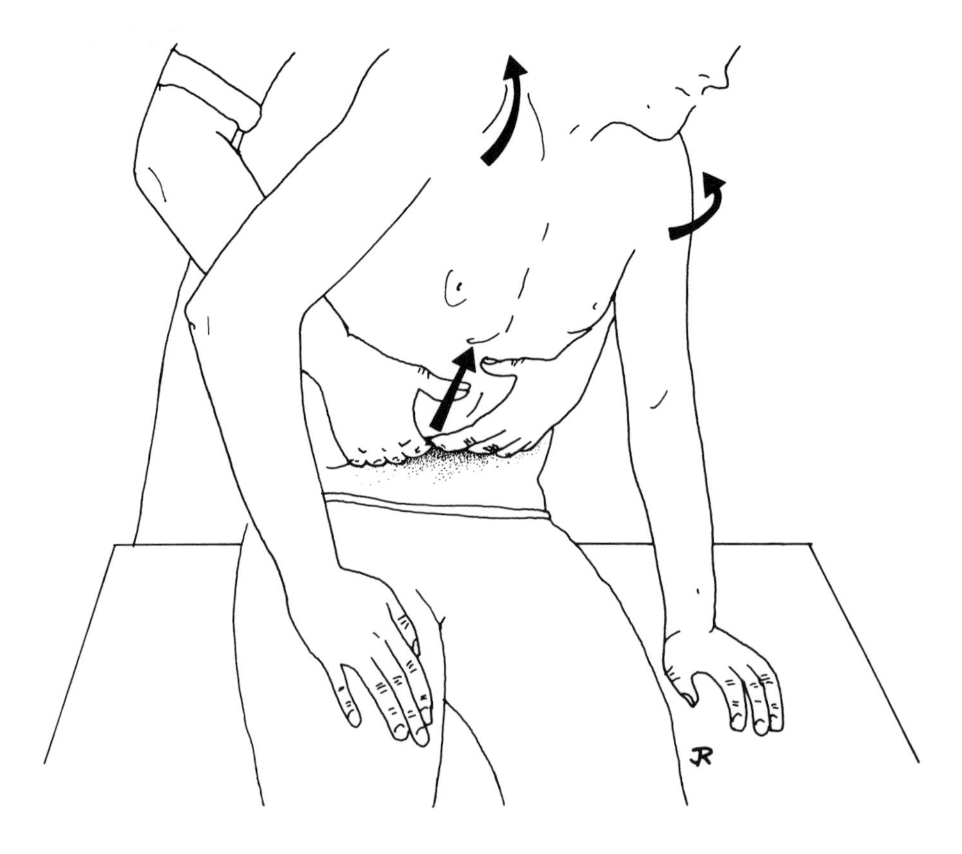

Illustration 8-9
Combined Manipulation of the Right Kidney — Seated Position

Another combined technique is used to stretch the ureters. This is particularly useful in cases of stones or ureterovesical reflux. Put the patient in the kyphosed seated position and place your hand firmly over the lower pole of the kidney. As you keep the kidney relatively fixed, extend the patient's spine and slightly rotate him toward the side being worked on. Repeat rhythmically until a release is appreciated. For problems at the ureterovesical junction, the technique is the same except that the hands are placed over the lateral angles of the bladder. This is a powerful technique and requires a very highly educated sense of touch.

Supine Position

With the patient supine, stand on the side of the affected kidney. Support the kidney with one hand and the knee with the other. Ask the patient to flex his hip against resistance, to contract the psoas and bring the kidney closer to the surface. Do this rhythmically and slowly (5-10 times per minute) and, with every relaxation of the psoas, push the kidney superiorly. This is very effective for mobilizing the kidneys, especially when there are adhesions between the renal fascia and the psoas. The technique can also be performed in the prone and elbow/knee positions.

Reverse Trendelenburg Position

The patient sits at the end of the table with hips and knees flexed and feet on the table. Sitting behind him on a stool, ask him to lean backward until his shoulders rest on your knees. His shoulders should then be approximately 40cm below the pelvis. Find the inferior pole of the kidney to be treated and then ask the patient to bend the ipsilateral leg toward the chest to tighten the psoas and relax the anterior abdominal wall. When you feel the inferior pole of the kidney, extend the hip and knees slightly to stretch the psoas. As you resist any further descent of the kidneys, they are mobilized via the psoas *(Illustration 8-10)*.

The basis of this technique is the same as in the direct technique, but by mobilizing the lower limb you induce the lumbar column to make a rhythmic rotational movement (clockwise or counterclockwise, depending on the side), which brings the kidney being treated forward each time. During each exhalation, your hand "lifts" the kidney.

We would like to reemphasize a very important point here. We have described above how, in case of a ptosis, the kidney usually slides downward along the psoas muscle and externally rotates in the frontal plane (the superior pole adducts and the inferior pole abducts), although the opposite occurs in very severe ptosis. Even if it is relatively easy to correct the actual ptosis, it is much more difficult to reestablish the proper axis of rotation of the kidney by making it rotate clockwise.

For this, with each of the techniques described above, try to contact the kidney on its inferomedial edge. If you cannot achieve this, be sure to reinstate the proper pendular motion using either induction or the following technique.

The best technique for correcting frontal rotation of the kidney is performed in the seated position. Simultaneously with the direct technique, have the patient rotate and sidebend the trunk to the side opposite the kidney being treated. This "double" movement moves the kidney forward (which helps treat the ptosis) and corrects the pathological rotation, as the inferior pole is abducted and the superior pole adducted. With your

Illustration 8-10
Combined Manipulation of the Left Kidney —
Reverse Trendelenburg Position

hands pressing on the kidney, repeat this rhythmically until a release is felt. This technique is more powerful than induction but less precise.

INDUCTION TECHNIQUES

Although it may seem fanciful to attempt to induce a motion across 15cm of tissue in a controlled manner, our experience has shown that this is in fact possible. All manipulative treatments of the kidneys should finish with induction, followed by a recheck with "listening."

Place the patient in the supine position. Place your hand longitudinally with the heel one finger's width from the umbilicus and the ulnar side of the hand oriented laterally on the abdomen. Encourage the part of the cycle with the greatest ease of motion and passively follow the other part of the cycle until a release occurs, with or without first passing through a still point *(Illustration 8-11).*

Always follow the same protocol: go from "listening" (diagnosis) to induction (treatment), and then "listening" (recheck). There should be a harmonious rhythm with a frequency of approximately seven cycles per minute. By "harmonious," we mean that there

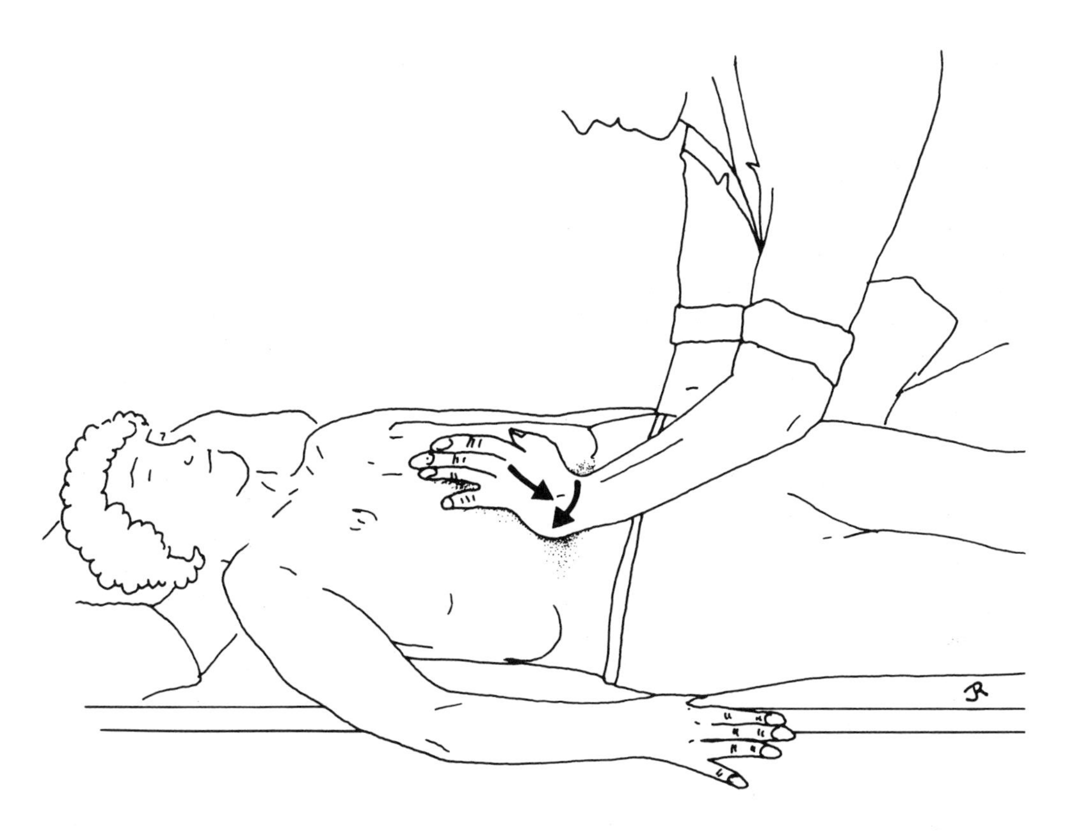

Illustration 8-11
Induction of the Right Kidney — Inspir

should be an even balance between the vertical, rotational and pendular movements in both inspir and expir. Normal motility was described above (page 194).

Adjunctive Considerations

ASSOCIATED OSSEOUS RESTRICTIONS

The most common vertebral restrictions associated with kidney dysfunction are T11/12 and their costovertebral joints. The entire region from T10 through L1 may be involved. Neurological reflexes may involve T6/7 and L1/2. When there is a dysfunction of the kidneys, always treat the kidneys first before considering vertebral manipulations.

Mechanical problems of the thoracolumbar junction affecting kidney function are quite common and easy to understand. Dysfunctions of the lower ribs often adversely affect the kidneys as well. There also seems to be a link between the kidneys and the coccyx which is less easy to explain; coccygeal restriction and renal ptosis often occur together.

Disturbances of kidney motion may be due to distant malfunctions. For example, articular restrictions may be found at the level of the superior tibiofibular joint or the cuboid articulation. We believe there is a muscular chain whereby the tibia is linked to the psoas muscle.

RECOMMENDATIONS

Be patient in your diagnosis; do not jump to conclusions in attempting to link the observed symptoms to suspected musculoskeletal or visceral causes. We believe that there is a hierarchy among tissues, and that viscera have significant primary dysfunctions much more commonly than do musculoskeletal structures. Remember that too much force or inappropriate directions of force can easily irritate visceral tissues. As you are working deeply when treating the kidneys, you must be especially careful. We caution you again not to disturb the iliac arteries and abdominal aorta when you are looking for a renal ptosis.

We recommend to our patients an exercise in the reverse Trendelenburg position with the pelvis placed higher than the trunk, on a sofa or armchair. The patient emphasizes exhalation, possibly by pulling a belt upward around the abdomen during exhalation. We recommend about twenty repetitions, advising caution in hypertensive patients. This technique should be practiced on an empty stomach, ideally just before mealtime, to avoid short-circuiting digestion or increasing vasomotor congestion caused by digestion.

Advise a patient to be careful when coughing, sneezing, etc., because such actions undeniably favor renal sliding when the internal ligaments are lax. Suggest that, if in a standing position, he immediately push his hands against a table or his thighs to stabilize the trunk.

If a patient has known calculi, warn him of the risk of their migration. He must significantly increase fluid consumption in order to flush out the urinary tract. It is common to find migrations of stones or very small calculi after our manipulations. Lemon juice added to water increases the efficacy of our manipulations on the kidneys and urinary tracts.

We do not recommend trusses or support belts because they compress the abdomen — they are useful while in position, but once they are taken off the symptoms reappear. The rare cases when trusses may appear useful are in thin female patients who are totally hypotonic, with kidneys descended almost into the pelvis and where you cannot count on normal physiological reactions to take over from your manipulations. In such cases the truss should be placed down low. Actually, results from this treatment are purely palliative, since no cure is taking place. We emphasize again that the motion of the kidneys is more important than their position.

Clinical Example

This case is remarkable more because of the symptoms presented than because of the results — we were not able to stabilize the patient's condition.

Mrs. X, twenty-eight years old, presented after the first period postpartum with an arterial tension of 220/113mmHg, diagnosed as idiopathic hypertension. Clinical examination of this overweight woman revealed nothing except for an increase in temperature over the surface projection of the left kidney. The requested urinalysis was negative. Radiography indicated a left renal ptosis of approximately 5cm. There were no major vertebral restrictions.

Manipulation was performed to improve the mobility of the left kidney and raise it, followed by a short session of induction. Following treatment, motility was good, and eight minutes later her blood pressure had fallen to 140/80mmHg. Unfortunately, circumstances prevent us from seeing this patient more than once a year and we are unable to stabilize this kidney. By stabilize, we mean maintain its motility and not necessarily its position. But, apart from an annual crisis, she feels well and lives normally — which wasn't the case before treatment. When we mobilize a kidney, improvement is due not to a new position (because it doesn't last), but to a stimulation, dynamization and revitalization of the organism.

Chapter Nine: The Perineum and Bladder

Table of Contents

CHAPTER NINE

The Perineum and Bladder

The perineum is a group of soft tissues which shut the pelvic outlet off from the outside. It has two conflicting roles: to form a solid and elastic floor and to open periodically to allow expulsions.

The abdominal viscera constitute a homogeneous column which rests on the internal iliac fossae and the superior edge of the ischiopubic rami. Obviously, some pressure is exerted on the pelvic inlet. In the case of the pelvic organs, the force of gravity predominates over the force of thoracic inhalation, and we cannot say that the organs are suspended from the diaphragm.

The pelvic organs rest upon the perineal floor, which must be closed in order to prevent a ptosis. If this floor were rigid, the organs would be compressed between it and the pressure of the visceral column. The floor must be able to alleviate these permanent pressures and also to compensate for the transient or temporary increased pressures produced by coughing, sneezing, hiccups, pregnancy, etc. In other words, it has to be elastic.

Within this floor are the openings for the urogenital tract and the rectum. There are three orifices in women and two in men. These orifices are weak points and, because of their number and size in women, the female perineum is by far the more fragile. The striated muscles which make up the perineum have an apparently contradictory double role. They have to seal the pelvic outlet and also be able to selectively open the individual orifices. To permit this double function, the muscles are numerous, varied and perpetually "hypertonic"; it is actually the relaxation of specific muscles which causes the orifices to open.

The bladder is a muscular/membranous reservoir which must store the urine and then contract to expel it via the urethra. The urge to urinate is normally felt when the volume of urine reaches 350ml. Incontinence (the inability to sense this urge or to control reflex evacuation of the bladder) is much more common in women, particularly post-partum women. This condition is not really an illness, but is a great social handicap. We have treated many women upset by incontinence problems who thought that surgery was the only possible solution. Their confidence undermined, such patients subconsciously feel like a child who cannot help wetting herself. Helping them is extremely worthwhile and satisfying.

Anatomy

PERINEUM

The muscles of the perineum form a net strung between the sacrum, coccyx, ischial tuberosities and pubis. Each muscle has a specific role as support, sphincter or both. The lateral walls consist of striated muscles, the obturator foramen is closed by the internal obturator muscle, and the space between the greater sciatic notch and the sacrum is closed by the piriformis muscle.

The floor is made up of the levator ani, perineal transverse and sphincter muscles. The levator ani muscle forms a lozenge-shape bounded posteriorly by the coccyx and two small sacrotuberous ligaments, and anteriorly by the pubic symphysis and two tendinous arcades distributed between the ischial tuberosities and symphysis *(Illustration 9-1).* This

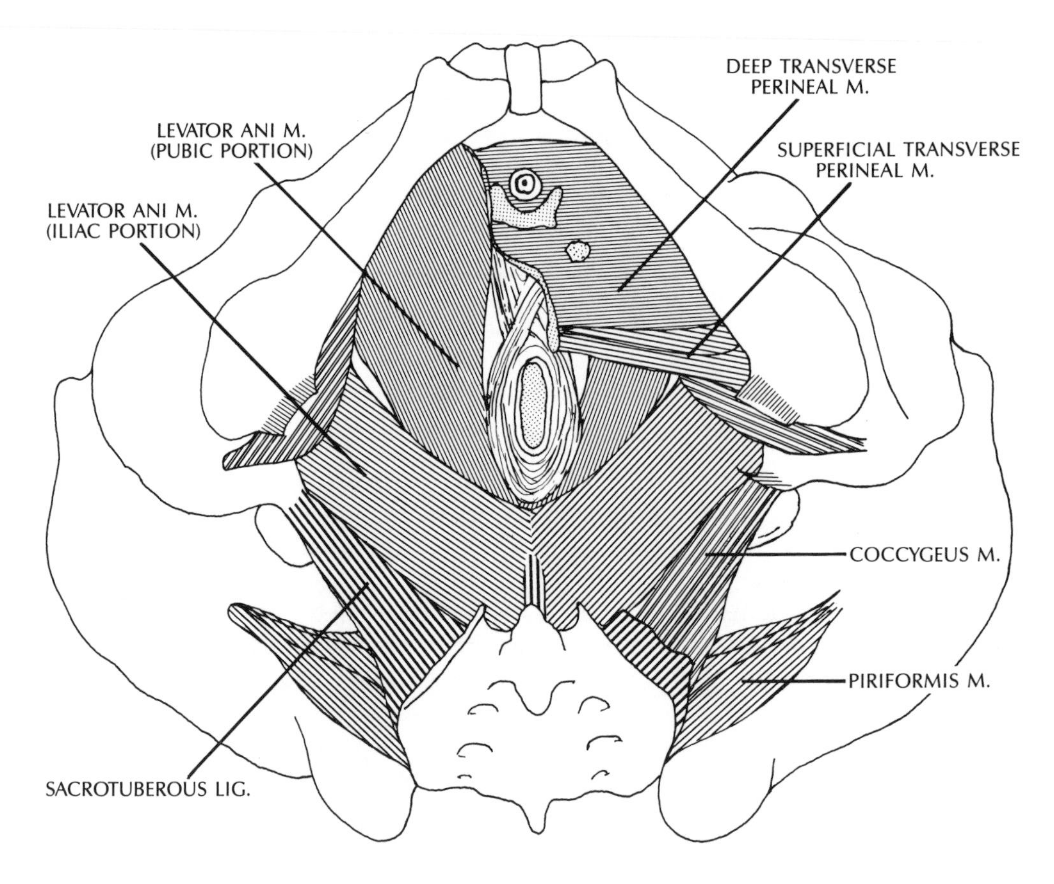

Illustration 9-1
Inferior View of the Female Pelvic Floor — Deep Muscles

muscle inserts on the pubis and tendinous arcade by fleshy fibers which pass around the rectum to insert on the opposite arcade. The fibers have a semicircular direction, the most posterior being inserted in the coccyx. The levator ani can be divided into a pubic portion (also called the pubococcygeus) and the more lateral iliac portion (also called the iliococcygeus). Another muscle, the coccygeus, which connects the ischial spine to the coccyx and sacrum, is continuous posteriorly with the levator ani and lies in the same plane.

The transverse muscle of the perineum closes the ischiopubic space. It is formed by the deep transverse muscle at the front and back and, more superficially, by the transverse superficial perineal muscle.

The sphincter muscles are the most superficial. The bulbospongiosus muscles are sagittal and in males are joined together by a median raphe. The ischiocavernous muscles are lateral and parallel to the ischiopubic rami. The external anal sphincter muscle is in a sagittal plane, located between the anococcygeal muscle and the raphe of the bulbospongiosus muscle in males. In females, the fibers cross in front of the anus on the tendinous center and thereby form the bulbospongiosus muscle *(Illustration 9-2).*

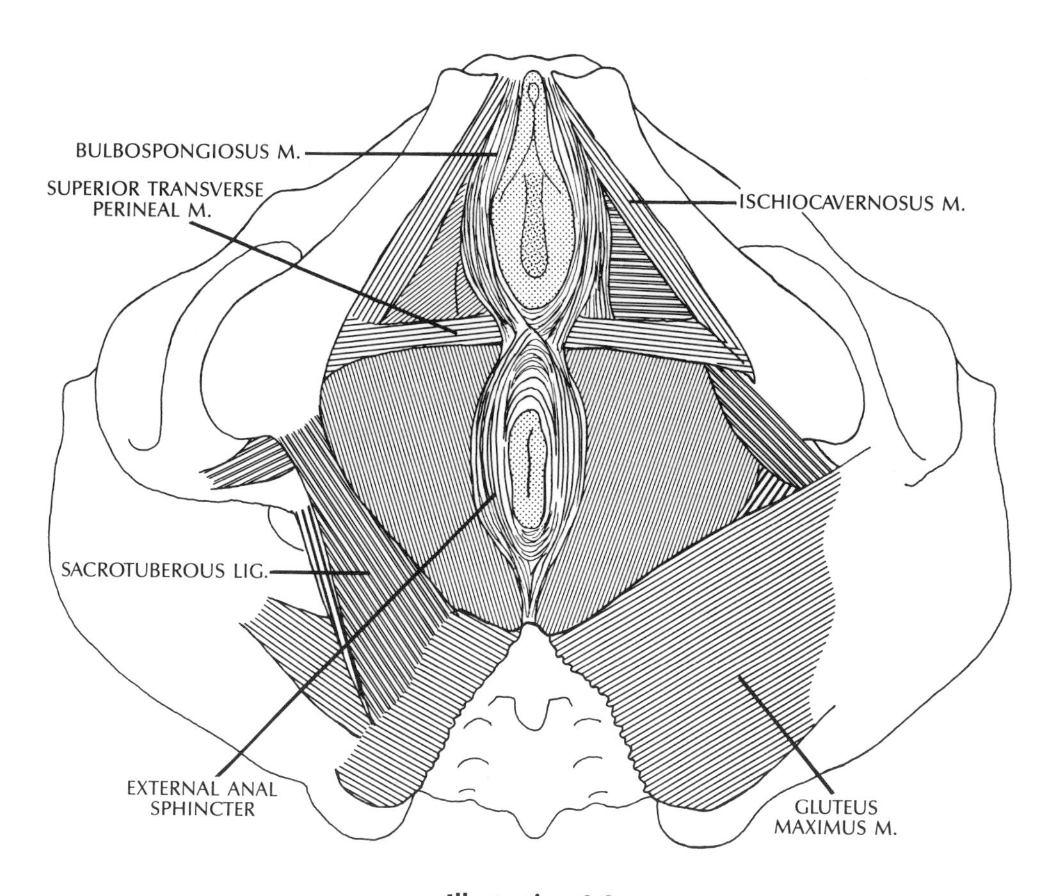

Illustration 9-2
Inferior View of the Female Pelvic Floor — Superficial Muscles

All the muscles which take the form of shoelaces partly or totally surrounding an orifice are sphincter muscles. The muscles with parallel fibers not diverted by an orifice, such as the iliococcygeus and coccygeus, have no sphincter role and serve to support the pelvic viscera. The exception is the deep transverse perineal muscle which is a flat muscle, crossed by the ureter, which acts as a sphincter. Actually, this muscle sends circular or semilunar fibers around the ureter — thus we find the shoelace-shaped muscles again. In the standing position, this muscle helps support the pelvic viscera.

BLADDER

The form of the bladder varies according to its degree of fullness, and its relationships with surrounding organs are obviously different in males and females. The empty bladder is confined to the anterior pelvic cavity behind the pubic symphysis. The full bladder projects up to 3cm above the symphysis and occupies part of the anteroinferior abdominal cavity as well. This is its most vulnerable and accessible position.

In the male, the bladder is contained within a cavity determined by the pelvic organs and their surrounding tissues. It is separated from the cavity wall by a perivesical space filled up by loose connective tissue. The cavity is bounded by the pubovesical ligaments anteriorly, the prostatoperitoneal aponeurosis posteriorly, the internal obturator and levator ani muscles laterally, the prostate and pubovesical ligaments inferiorly and the peritoneum (which is raised by the bladder when full) superiorly. This arrangement results in several cul-de-sacs: a prevesical cul-de-sac, two lateral ones and a posterior one.

In the male, the bladder relates superiorly to the prevesical cul-de-sac, to which the median umbilical ligament is attached. This ligament (which runs along the anterior abdominal wall and connects the apex of the bladder to the umbilicus) is a remnant of the urachus, an epithelial structure which, in the embryo, connects the bladder to the allantois. Laterally, the bladder relates to the levator ani and internal obturator muscles, as well as to the sagittal peritoneal folds. Posteriorly, it relates to the small intestine, sigmoid colon and rectum. The bladder is covered by the peritoneum which forms Douglas' pouch. Inferiorly, the neck of the bladder is found about 3cm behind the middle part of the symphysis. The neck rests on and is in direct contact with the base of the prostate and the deferent ducts. The inferior bladder relates with the seminal vesicles and deferent ducts laterally, and medially to the triangular space between the two deferent ducts. It contacts the rectal ampule via the prostatoperitoneal aponeurosis.

In the female, the bladder is generally comparable with that of the male, but obviously differs in its relationships with the uterus, which are analogous to those with the seminal vesicles and prostate in the male. The upper third of the inferior bladder is in contact with the cervix of the uterus; the rest is connected to the vagina. A slack cellular tissue separates the uterus from the bladder. Between the bladder and vagina is the vesicovaginal partition; between the bladder and uterus, the vesicouterine cul-de-sac. The posterior bladder surrounds the uterine isthmus.

The small bevel-edged fissures which connect the urethral orifice to the two ureteral orifices run anteromedially. Their largest diameter is 3-5mm. When urine enters the bladder, they become round and full. The posterior urethral orifice is called the neck of the bladder and is a transverse fissure with two lips.

The peritoneum, which covers the superior aspect of the bladder, bends back upon itself and continues until it reaches the lateral walls. At this point, it runs posteriorly with the anterior layer of the large ligaments.

Visceral Articulations

The peritoneum covers the bladder and connects it to the rectum posteriorly, the abdominal wall anteriorly and the pelvic walls laterally. Therefore, whereas the vesical cavity is very strong inferiorly and laterally, it is easily depressed at the top. The bladder is separated from surrounding structures by a slack cellular tissue, reinforced to form the umbilicovesical aponeurosis anteriorly, the prostatic portion of the sagittal peritoneal folds laterally and the prostatoperitoneal aponeurosis posteriorly. The vesical cavity lines a larger serous cavity, which is subdivided into prevesical and retrovesical spaces.

The prevesical space (also called Retzius' space) extends from the pelvic floor to the umbilicus. Anteriorly, it is limited by the pelvic abdominal wall, pubovesical ligaments and the pubis. Posteriorly, it is bounded by the umbilicovesical aponeurosis, which arises from the umbilicus, goes in front of the median umbilical ligament and inserts on the apex of the bladder. This aponeurosis then enlarges to surround the anterior and lateral sides of the bladder, and blends into the pubovesical ligaments, pelvic aponeurosis and aponeurosis of the internal obturator muscle. The pubovesical ligaments contain muscular fibers and can exchange fibers with the bladder itself. In sagittal section, the prevesical space appears as a triangle of which the umbilicus is the summit and the pubovesical ligament the base *(Illustration 9-3).*

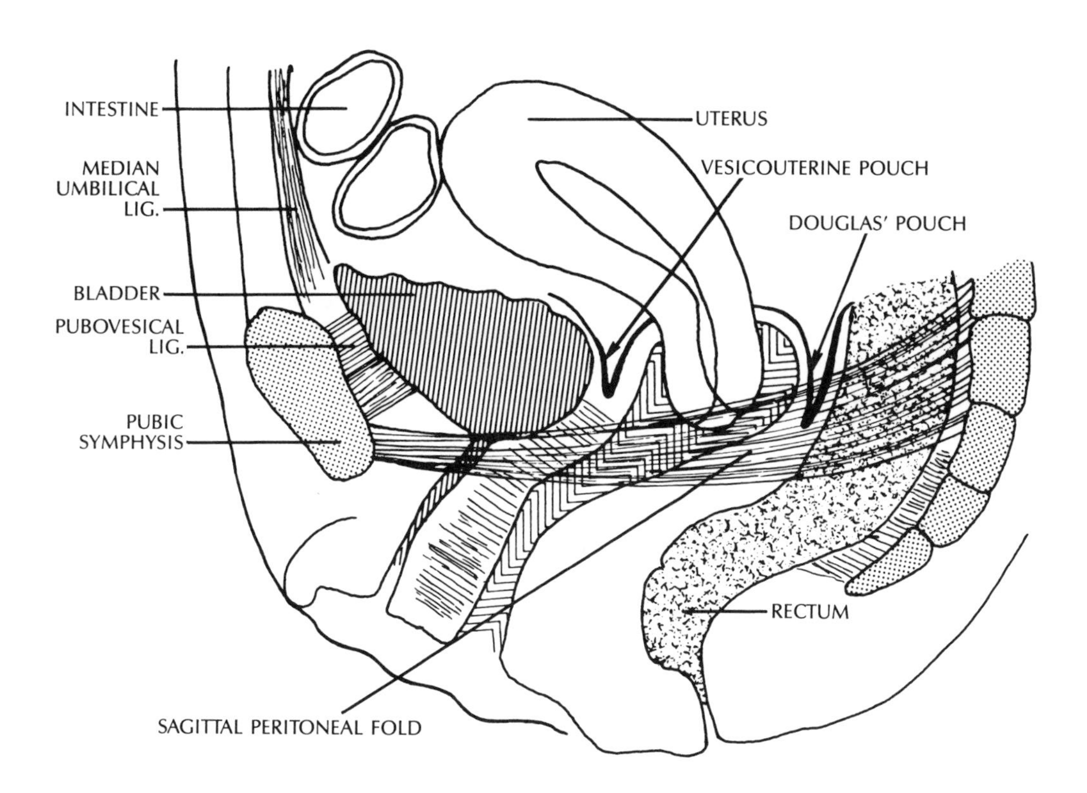

Illustration 9-3
Visceral Articulations of the Bladder in the Female

The median umbilical ligament is accompanied by two fibrous cords (called, confusingly, the *medial* umbilical ligaments), which are just lateral to it and are the remains of the umbilical arteries. The other major ligaments are the pubovesical ligaments, which solidly attach the anterior and inferior bladder to the pubis. Local adhesions can connect the base of the bladder to the prostate and the pelvic floor; it is the pelvic floor which plays the greatest role in restrictions of the bladder, via the prostate.

Intracavitary pressure and turgor continue to play a role in this region of the body. Each pelvic organ has a contiguous relationship with the peritoneum, and is affected by intracavitary pressure via the intraperitoneal abdominal organs. However, these effects are felt less in this area, which may explain the relatively high frequency of ptosis of the pelvic organs.

Sliding Surfaces

Via the peritoneum or the aponeuroses, the bladder articulates with the small intestine, uterus and rectum. As explained below, the bladder is mostly affected by positional problems, injury to the sliding surfaces affecting it less than in the case of the abdominal viscera.

TOPOGRAPHICAL ANATOMY

The apex of the bladder (and the prevesical space in front of it) is normally found at the level of the pubic symphysis, and rises as the bladder fills up. The superior insertions of the median and medial umbilical ligaments occur at the inferior edge of the umbilicus.

Physiologic Motion

The movement of the diaphragm affects the bladder much less than it does the abdominal viscera. Basically, the bladder moves in sync with the sacrum and uterus, i.e., posterosuperiorly with inhalation and anteroinferiorly with exhalation. Most functional problems of the bladder occur following a ptosis which modifies the action of the sphincters and inhibits proper physiology. The following section will discuss the muscles of the bladder and their role as sphincters.

VESICAL MUSCULATURE

The lower third of each ureter is accompanied by an external muscular sheath which surrounds the ureteral ostium like a shoelace. As the bladder fills, the ostium moves upward and opens. Between the two ostia are the muscular "laces" which pull the ostia downward and close them *(Illustration 9-4)*. In cases of vesical ptosis, the ostia close, leaving urine to stagnate in the ureters with inherent risks of secondary infection.

Similar shoelace-like muscle fibers surround the anterior circumference of the internal urethral ostium. They pull the anterior wall toward the back to close the ostium, forming the involuntary sphincter of the bladder. Closing of this sphincter is assisted by fibers of the pubovesical muscle (which surrounds the posterior circumference of the ostium) and the trigone (the smooth triangular surface on the inner vesical wall bounded by the openings of the two ureters and the internal urethra).

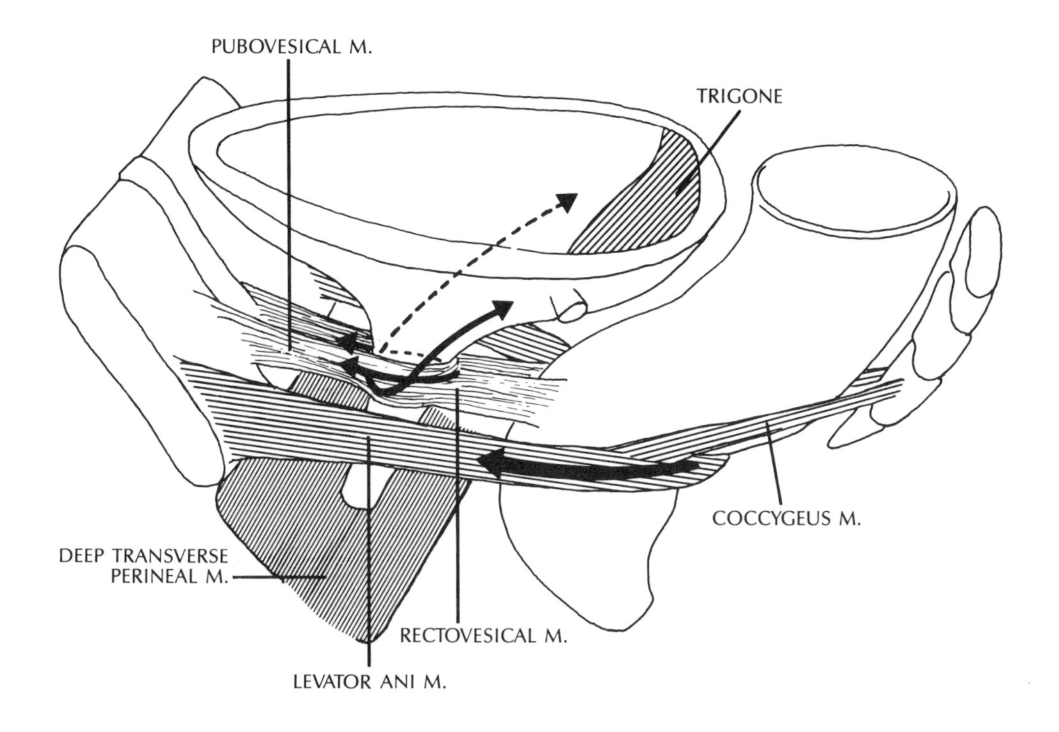

Illustration 9-4
Physiology of the Vesical Ostia

During active involuntary opening of the urethra, the occlusion loop of the ostium comes from the muscular fibers in the sphincter. As they contract, they form a furrow on the floor of the bladder which helps to open the urethra. This function is reinforced by the fibers of the rectovesical muscle located behind the pubovesical muscle. The "voluntary sphincter" comprises the deep perineal transverse muscle (crossed by the urethra, which is encircled by spiral loops), levator ani and coccygeus.

The wall of the bladder is made up of different fasciculi. The external fasciculi arise anteriorly to the neck of the bladder and pubovesical muscle, and are distributed toward the apex. The pubovesical muscle sends some fibers onto the median and medial umbilical ligaments, prostate or anterior vagina.

URETHROVESICAL DISPLACEMENT

The principal problems of the bladder result from collapse of the supporting structures. In women, this collapse often happens after a baby has been delivered by suction, or through an episiotomy which is too large. The perineum then loses much of its contractility and elasticity, which diminishes its sphincter role. Other causes of collapse are old age, depression and ptosis of an abdominal organ which puts weight upon the bladder (e.g., intestinal ptosis, anteversion of the uterus, impacted feces in the sigmoid colon). In general, anything which is capable of pushing the bladder and pelvic floor downward can lead to collapse. We shall discuss displacement of the coccyx in chapter 11.

Huguier and Bethoux (1965) performed experiments on the consequences of urethrovesical displacement, using intravesical and intraurethral recordings. Their study is important because it shows the wide-ranging effects of displacement of organs, even if minimal. When the urethral sphincter is in the abdominal cavity in a continent woman, abdominal pressure is exerted upon the bladder and proximal urethra, strengthening them. If the proximal urethra moves out of the abdominal enclosure following a perirenal collapse, abdominal pressure will no longer reinforce the urethral sphincter; on the contrary, it will increase pressure within the bladder and provoke incontinence *(Illustration 9-5)*. These conclusions can be generalized to other sphincters of the body and give you a better idea of the dysfunctions which can be produced by visceral restrictions and ptoses.

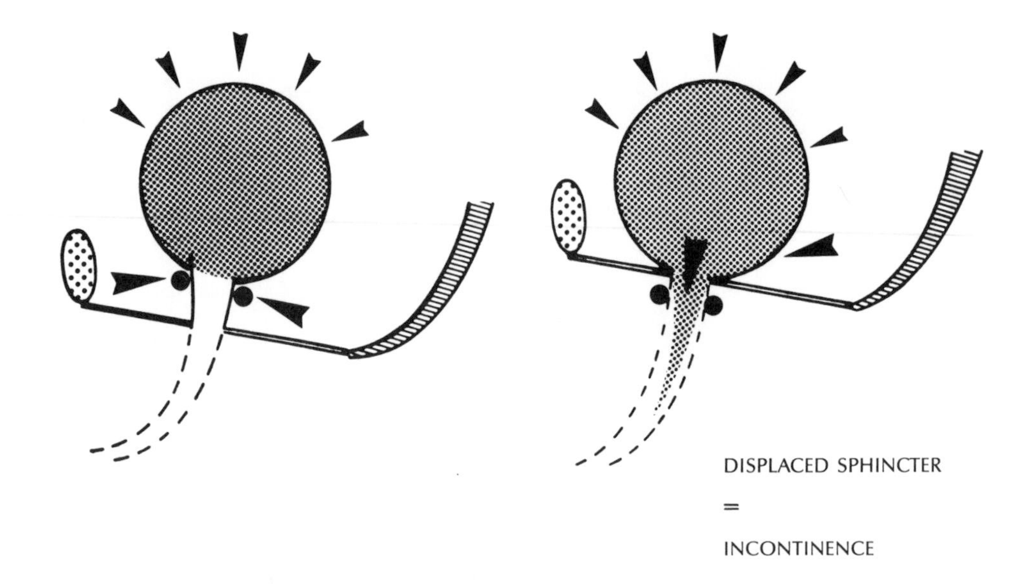

Illustration 9-5
Pathophysiology of Urethro-vesicular Collapse

This illustrates the important role played by the perineum. Abdominal pressure on the pelvic cavity is contained anteriorly by the abdominal wall, posteriorly by the coccyx and inferiorly by the perineum. Imagine a patient who falls on the coccyx; it will be moved into a position of anterior flexion. The sacrococcygeal is one of the rare articulations (along with the sternoclavicular and chondrocostals) which may undergo displacements (positional restrictions). The anterior flexion position will relax the perineal fibers and lower the proximal urethral sphincter. This partially explains incontinence following falls onto the coccyx. In the next chapter we shall discuss other effects of this injury on the uterus. The perineum must have good tone and be stretched in order to be function properly. If the coccyx moves closer to the pubic symphysis, the fibromuscular elements will do the same and automatically lose part of their contractility and tone, thereby losing their ability to support the sphincter (see chapter 11).

MOTILITY

The bladder and uterus show great similarity in motility. During expir, the apex of the bladder moves first posteriorly, then superiorly, as if it wished to reunite with the umbilicus.

Indications for Visceral Manipulation

Indications are mostly related to mechanical troubles of the bladder. They include infections from ureterovesical reflux, which can have dangerous consequences for the kidneys (pyelonephritis), as well as those from urinary stasis, usually from recurrent *E. coli* infections. Visceral manipulation of the bladder is particularly effective for incontinence, and should also be considered in cases of uterine malpositioning, dyspareunia linked with bladder problems and ptosis and restrictions of the bladder and surrounding organs.

We have our patients with recurrent urinary tract infections checked by a urologist first for strictures, structural problems of the urinary tract or severe infections. If these are not present, we have found manipulation effective for this problem. We have stated in other chapters that acute infections are a contraindication for visceral manipulation. However, this does not apply to most lower urinary tract infections, primarily because there is no risk of irritating the peritoneum by manipulating the bladder.

Evaluation

A complete history is essential because it allows precision as to the kind, quality and intensity of the problems felt — incontinence, polyuria, heaviness in the lower abdomen, color of the urine, etc. In the case of hematuria, it is essential to find the cause by means of ultrasound or appropriate radiological exams. If the feeling of heaviness in the lower abdomen is lessened when the patient holds up her abdomen, there is either a vesical ptosis or uterine problem, the mechanical pathophysiologies of these two organs being closely related. It may also be useful to ask if: (1) incontinence only occurs during effort, coughing or constipation (i.e., is stress incontinence); (2) it followed childbirth, surgery or a fall on the coccyx; or (3) sexual intercourse provokes an urge to urinate.

There are other important tests. Palpation and percussion are more difficult to analyze. It is easier to palpate the bladder when it is full, but unfortunately this provokes an urge to urinate which makes the patient ask you to stop the technique. If appropriate, obtain urinalysis for bacteria and blood cells.

MOBILITY TESTS

Mobility is tested by raising the apex of the bladder (via the median and medial umbilical ligaments) in order to appreciate its course and elasticity. This test is an important part of the exam. You can locate the apex of the bladder by placing your fingers on the superior edge of the pubic symphysis (slightly laterally), then pushing them posteriorly and finally superiorly. Make sure that the patient has urinated before beginning this technique. The patient should either be seated (so that the bladder is at its heaviest and stretches its ligaments) or supine with the hips and knees flexed. In these positions, the tension of the abdominal musculature is diminished *(Illustration 9-6).*

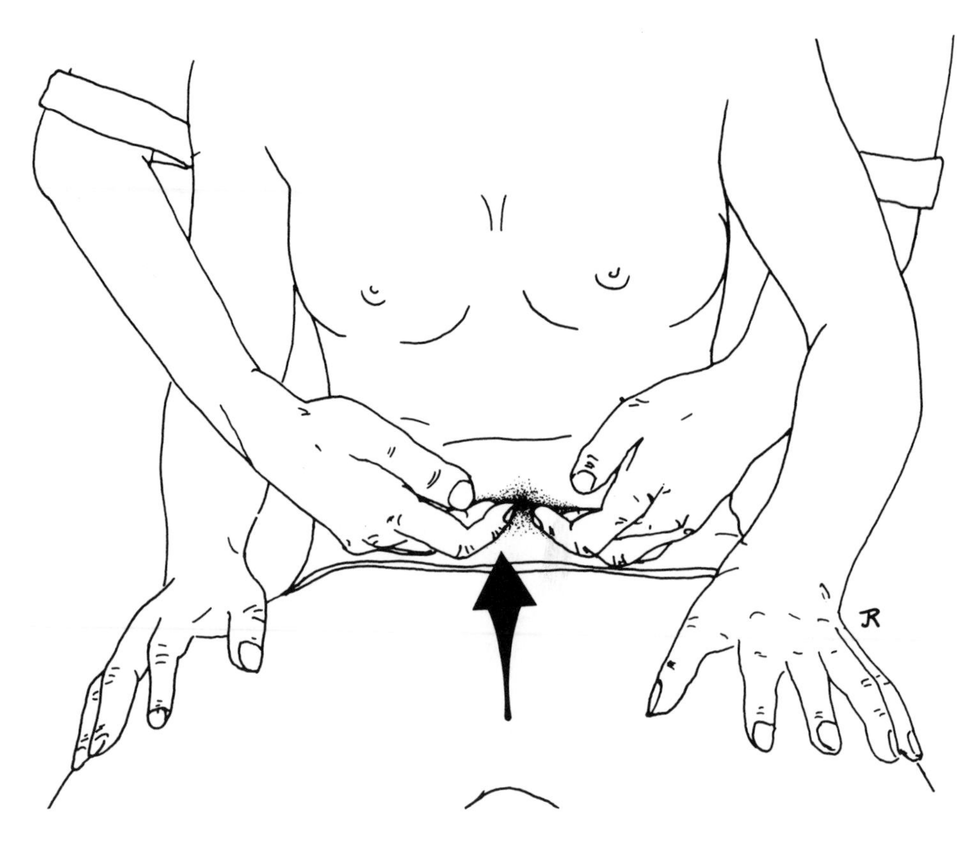

Illustration 9-6
Mobility Test of the Bladder — Seated Position

You can also utilize pressure points between the pubic bones and the umbilicus as a focal point from which to carry out this technique. These subumbilical points are particularly useful in the presence of a scar in the lower abdomen, as they tell you if there is abnormal tension above or below the scar. In the case of a marked ptosis, the patient should feel an uplift and lightening of the bladder region during the technique. It is imperative that the test for the coccyx (described in chapter 11) be associated with this one. In fact, the coccyx should be tested with any urogenital dysfunction.

In order to differentiate between a vesical and uterine ptosis, push the bladder toward the perineum with the patient in the seated position. If there is a vesical ptosis, this movement should trigger an urge to urinate and a feeling of weight. The feeling of weight is felt in the anterior perineal region, whereas uterine ptosis causes a more posterior tension.

MOTILITY TEST

With the patient in the supine position with legs bent, place the palm of your hand directly above the pubic symphysis with the fingers pointed toward the umbilicus. During expir, your hand should move posterosuperiorly toward the umbilicus *(Illustration 9-7).*

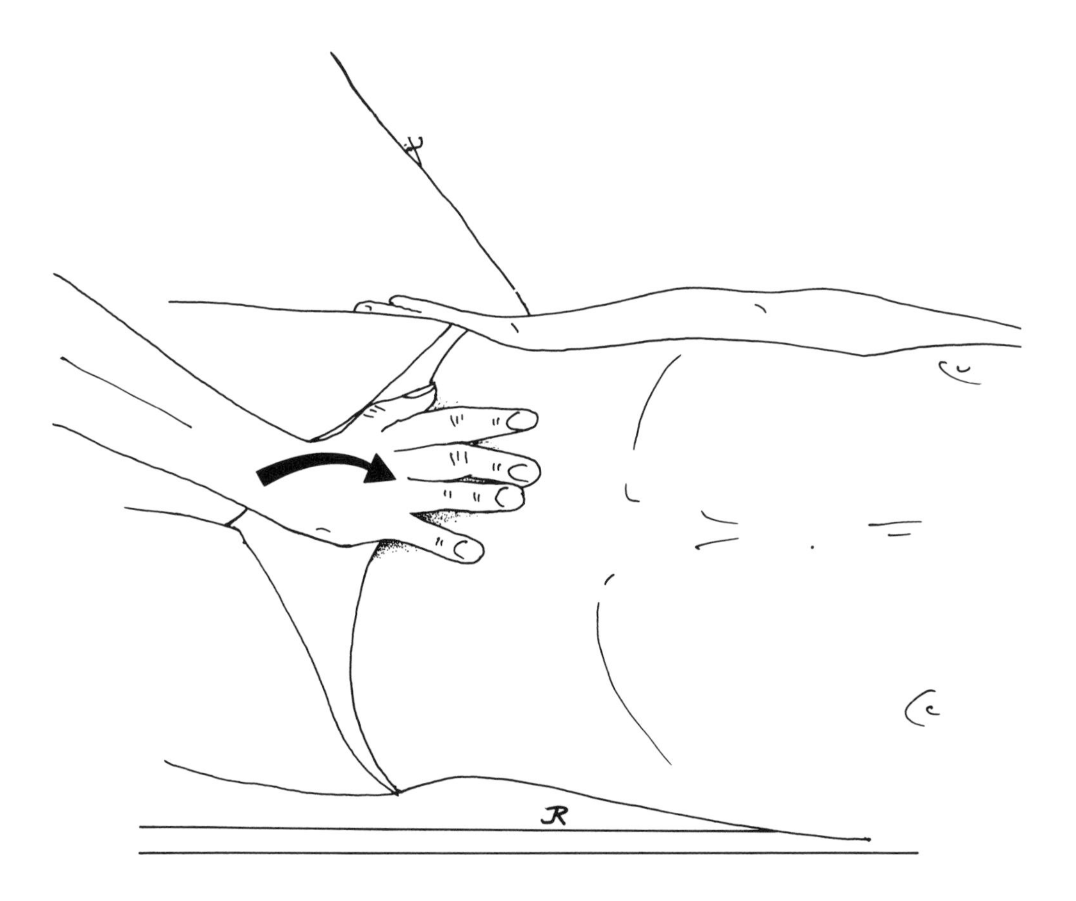

Illustration 9-7
Motility Test of the Bladder — Supine Position

Restrictions

There are several causes of restrictions and vesical ptosis:

- superiorly, the mass of the small intestine or even the stomach, which can force the bladder downward
- posteriorly, the uterus (or the prostate in the male), which can push the bladder downward and forward or pull it backward, depending upon its position and the tension in its surrounding ligaments
- anteriorly, a retraction of the pubovesical or puboprostatic ligaments that prevents the bladder from expanding upward as it fills; this automatically forces it downward and toward the perineum
- inferiorly, collapse of the perineal floor, leading to urethrovesical displacement

Manipulations

CONTRAINDICATIONS

There are two absolute contraindications to manipulation of the perineal region: presence of an IUD and pregnancy. Manipulation of the bladder with an IUD in place could injure the tissues and cause scars which can have dire consequences (e.g., sterility). Manipulations during pregnancy can lead to a miscarriage.

DIRECT TECHNIQUES

With the patient in the kyphosed seated position, place your fingers on the umbilicopubic line. Push the fingers posteriorly to create a fixed point and then straighten the patient up, which will increase the umbilicopubic distance and draw the vesical apex upward. Do this rhythmically several times, starting with your fingers just above the pubic symphysis and then moving them toward the umbilicus *(Illustration 9-8)*. At first,

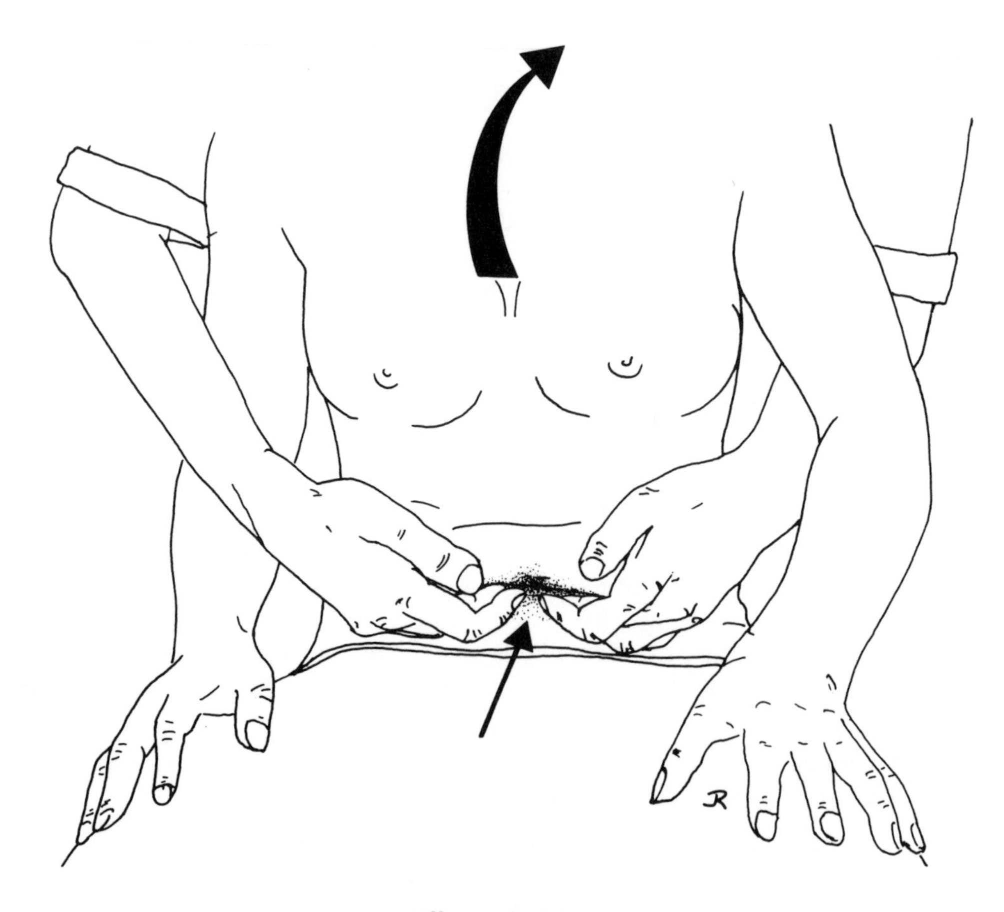

Illustration 9-8
Direct Manipulation of the Bladder — Seated Position

you will be acting only upon the pelvic organs, but as your fingers move upward, you will affect the greater omentum, small intestine, peritoneum, umbilicovesical aponeurosis, median and medial umbilical ligaments, via the anterior abdominal wall. You will automatically have an effect on the uterus, which covers the posterosuperior bladder (the most mobile part) and is obliged to follow it. In the male, this movement acts upon the prostate, which is closely connected to the bladder.

Alternatively, place the patient in the supine position with legs bent. Stand behind the patient's knees and place your fingers on the umbilicopubic line with an appropriate amount of pressure (enough to reach the median umbilical ligament but not the small intestine), and push the fingers posterosuperiorly toward the umbilical plane. For the technique to work, you must feel the sensation of a cord stretching *(Illustration 9-9)*. Relax your fingers and repeat the procedure in a slow steady rhythm of approximately one movement every ten seconds. When the tissues relax, usually after about ten repetitions, move your fingers closer to the umbilicus. When you are done with the median umbilical ligament, shift your fingers laterally so that the medial umbilical ligaments are more selectively stretched. Follow the tension of the tissues and use the same amount of pressure.

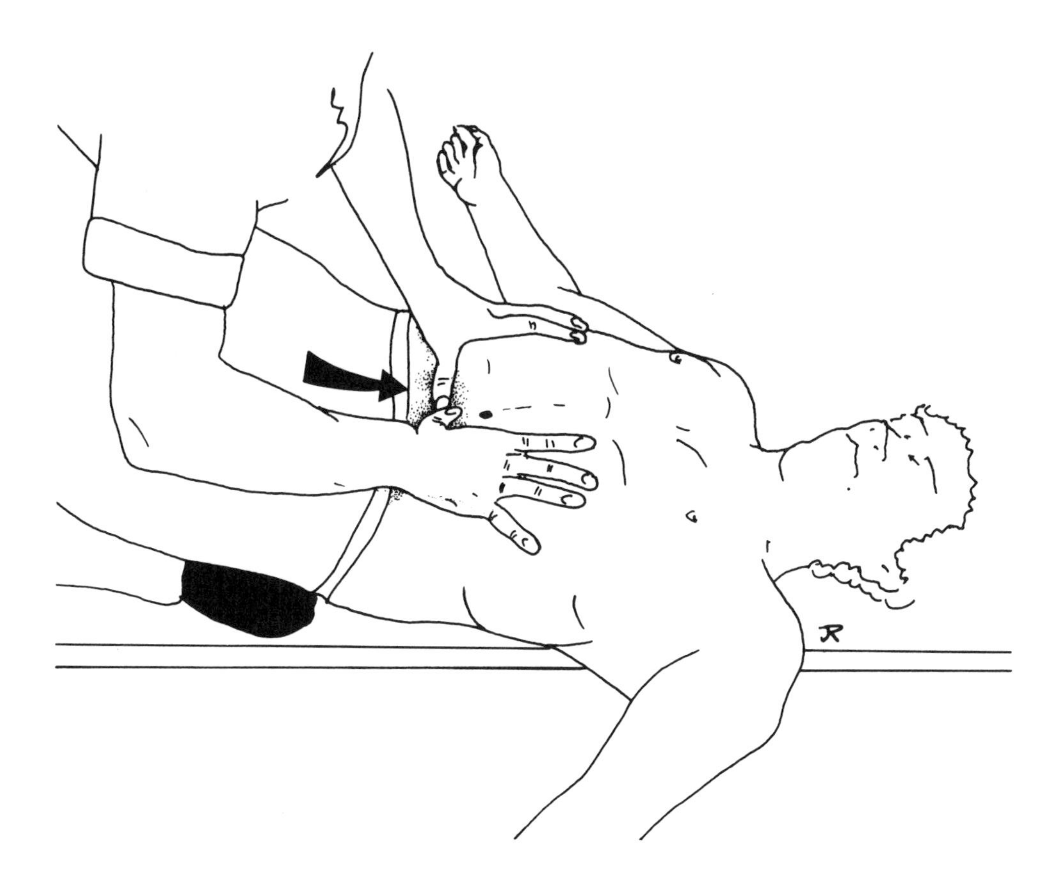

Illustration 9-9
Direct Manipulation of the Bladder — Supine Position

Stretching the pubovesical ligaments is also important, since bladder function will be affected if they are too tight. Because there can be an exchange of fibers between the bladder and these ligaments, a spasm of the muscle fibers in the ligaments can lead to frequent and urgent urination. With the patient supine with hips and knees flexed, stand at shoulder level on one side and put your fingers just lateral to the symphysis. After first pushing posteriorly, push along the underside of the pubic bones toward the genitalia. As the lateral aspects of the pubovesical ligaments release, move your fingers more medially until they meet when the ligament is fully released. This is a strong technique and may be quite uncomfortable. It is important to be firm but gentle.

You can exert some pressure on the inferior urogenital diaphragm by placing the fingers or thumbs either below the ischiopubic rami, or in the obturator foramen just lateral to the symphysis via the pectineus muscle. To determine if this technique is indicated, compare the tension in these two areas bilaterally. If one side is more tense, apply light pressure to it rhythmically until the tension releases. The technique utilizing the obturator foramen is more difficult, but more effective. We are indebted to our colleague, Sainte-Rose, for teaching us this procedure.

If the coccyx is restricted, it must be treated (see chapter 11).

COMBINED TECHNIQUES

With the patient in the seated position, you can combine direct pressure with sidebending of the trunk to more specifically manipulate the medial umbilical ligaments. If, for example, you sidebend the patient to the left, you increase tension on the apex of the bladder and the right medial umbilical ligament. There are numerous possible variations on this technique to localize your effects precisely where needed.

Alternatively, with the patient in the supine position with legs bent, place one hand on the umbilicopubic line and with the other mobilize the legs in a sidebending rotational movement so that you focus the traction on a particular area *(Illustration 9-10).* The more the legs are bent toward the chest, the more deeply you can apply pressure on the bladder and the more anterior restrictions of the bladder will be released. Rolling the knees to one side or the other increases the traction on the medial umbilical ligaments and makes it easier to mobilize the bladder laterally. This technique utilizes a combination of forces and can produce a quite specific and well-directed effect.

Combined techniques are also useful for treating the pubovesical ligament. With the patient supine and knees bent, perform the technique as described in the previous section on one side at a time. In addition, adduct and internally rotate the thigh on the side being worked on. This will make it easier for you to work deeply.

INDUCTION TECHNIQUES

For direct anterior induction, the patient is placed supine with the knees bent. Put the heel of your hand over the pubic symphysis with the fingers pointed toward the umbilicus. Encourage the movement of the apex of the bladder, which goes slightly posteriorly and then superiorly in expir *(Illustration 9-11).* The difference between direct and induction techniques in the pelvis is slight because the normal amplitude of urogenital organ motion is much less than that of other organs. This is due, in part, to their distance from the diaphragm and to structural constraints. The reduced amplitude is reflected in both mobility and motility, and fixations (decrease in both types of motion) are quite

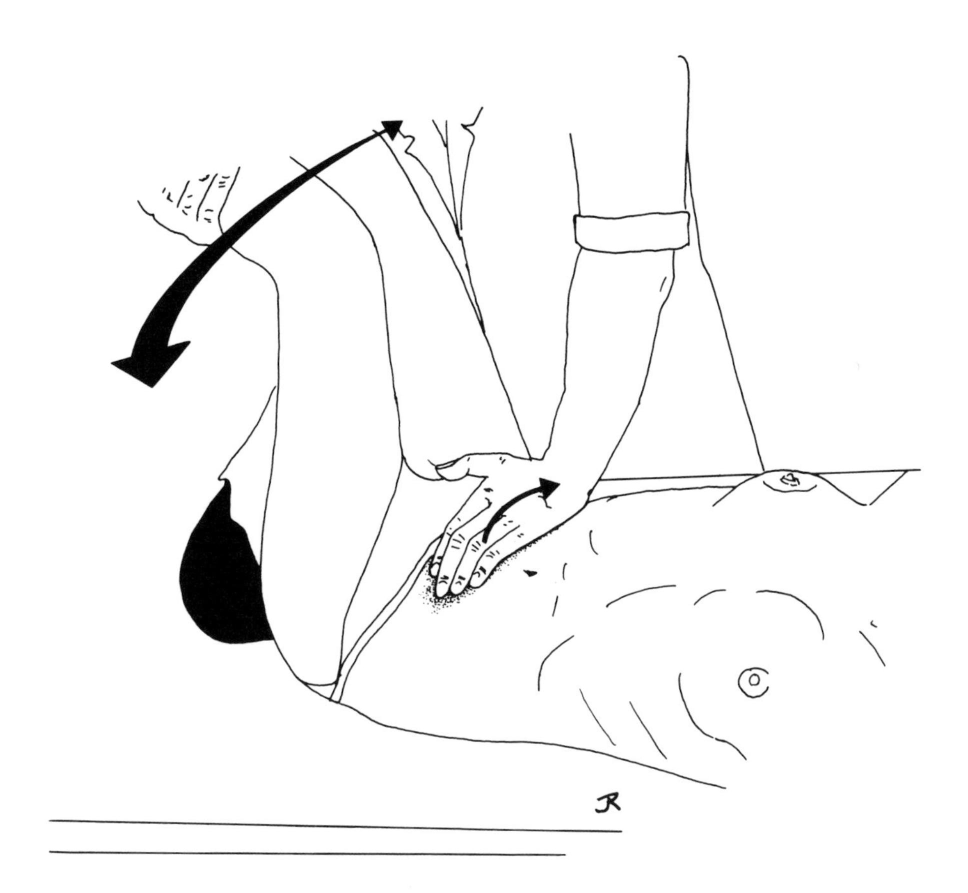

Illustration 9-10
Combined Manipulation of the Bladder — Supine Position

common. To be honest, the ability to perform induction in the pelvic region is usually acquired only after much practice, when the importance of precision over force is fully appreciated.

Direct anteroposterior induction is a combination of the direct anterior technique with induction of the sacrum. The anterior hand is placed as described above; the posterior hand cradles the sacrum in its palmar surface. Both hands work together; while the anterior hand moves posterosuperiorly, the posterior hand moves anteroinferiorly *(Illustration 9-12)*, and vice versa, until a release is felt (with or without a still point). The anterior hand acts upon the median umbilical ligament and the umbilicovesical and pubovesical ligaments, while the posterior hand works on the posteroinferior connections, including the rectovesical ligament, deep perineal transverse, levator ani and coccygeus muscles and sagittal peritoneal folds. This is a powerful technique that improves the most important motion of the urogenital area.

You can also combine sidebending rotations of the lower limbs with direct induction techniques. The difference between this and direct combined techniques is that in this one you follow the motion, whereas in the latter you direct the motion and consequently use more force.

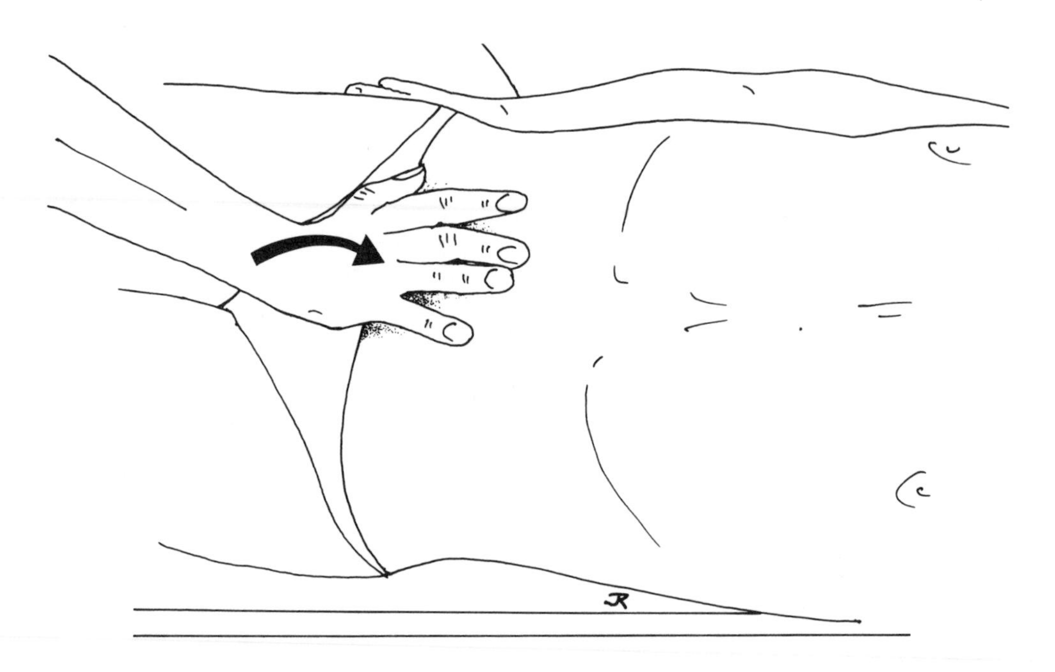

Illustration 9-11
Induction of the Bladder — Expir

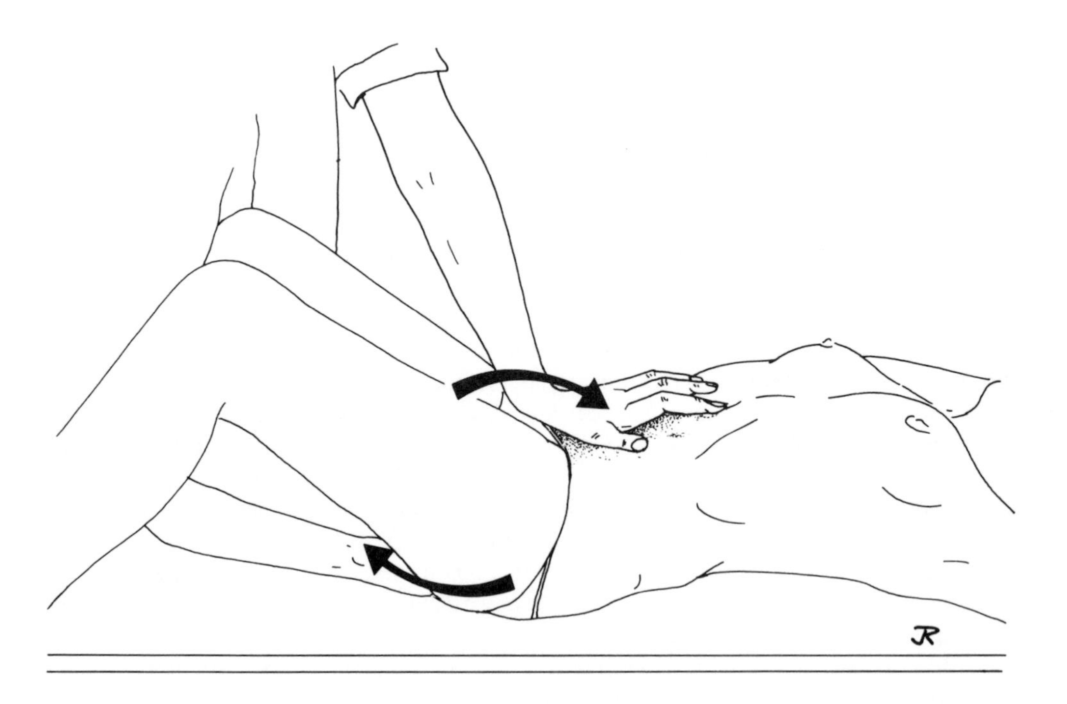

Illustration 9-12
Anterio-posterior Induction of the Bladder — Expir

Effects

All improvements of the mobility of the bladder will have beneficial effects on the vesical sphincters, on the prostate gland and uterus via their connections with the bladder and on urethrovesical displacement. We have been particularly successful with the latter problem, and have been able to show that a few millimeters of superior movement gained for the urethra and/or bladder enables them to recover their function entirely and thus avoid surgical intervention for the patient with incontinence. You should also be aware that the pelvis has a tendency to become congested because its venous circulation is often disturbed. Manipulation of the pelvic organs improves the circulation in this area.

When there are calculi present, it is imperative that induction techniques be used since these techniques are incapable of injuring the surrounding tissues. Calculi are not a contraindication to treatment, but do demand gentleness.

Adjunctive Considerations

ASSOCIATED OSSEOUS RESTRICTIONS

With bladder problems, one always finds restrictions of the sacrum, sacrococcygeal joint and feet. In our patients, we have frequently found restrictions of L2/3 associated with incontinence. Once again, remember that these restrictions are not necessarily an indication for spinal manipulation; your osteopathic knowledge will tell you if treatment of the vertebrae is necessary.

RECOMMENDATIONS

Be aware of the patient's menstrual cycle — efficacy will be increased when treatments are given within a few days after the end of the period. Premenstrual pelvic congestion makes these techniques cumbersome and difficult to perform properly. Always ask the patient to empty the bladder before treatment sessions. We often advise our patients to use the reverse Trendelenburg position (page 205) daily between periods. This technique is helpful for restrictions and ptosis of the bladder and particularly for incontinence. We also recommend it to patients with a history of heavy bleeding after childbirth, miscarriage or abortion, retroflexed uterus, pelvic congestion or scars. Some practitioners advise perineal gymnastics which, apart from stimulating the levator ani muscle, have not yet proved their worth to us.

Chapter Ten: The Female Reproductive System

Table of Contents

CHAPTER TEN

The Female Reproductive System

The male internal reproductive system consists of the seminal vesicles and other small structures which are almost entirely covered by the bladder. Only the bladder and rectum hold up the peritoneum, by forming Douglas' pouch. In contrast, females have an extensively developed internal reproductive system consisting of the vagina, uterus, ovaries and associated structures. Much of the uterus is separated from the subperitoneal tissue by an uplifting of the serous membrane which forms, on the left and right, broad ligaments. The uterus and these ligaments subdivide the pelvis into two cavities, preuterine and retrouterine. The close relationship between the peritoneum and uterus explains the frequency of uterine dysfunctions occurring secondary to peritoneal dysfunctions. Many of the complications associated with manipulative treatment in women arise from this phenomenon. Infections, pregnancy, ptoses, surgical interventions, "inevitable" episiotomies, suction in childbirth, etc., give rise to a variety of functional disorders for which osteopathic intervention is indicated.

Anatomy

The uterus is cone-shaped, flattened from front to back and divided into a body and cervix which are connected by the isthmus. In multiparous women, the uterus is 7-8cm long and 5cm wide at the base, and weighs 30-40g. It is of a fairly soft consistency and impressions are often left on it by the small intestine. Its capacity to expand is demonstrated by the fact that by the eighth month of pregnancy the top has risen to the level midway between the xiphoid and the umbilicus. The uterus is very mobile, and its position varies depending on whether the bladder and rectum are empty or full, and whether or not the intestinal loops penetrate the cul-de-sac. Usually, it is bent forward and tilted forward (anteverted and anteflexed).

The uterine tubes, also known as fallopian tubes, are looping cylindrical ducts 10-12cm long. Running through each tube is a canal which is 1mm in diameter medially and grows to 4-6mm laterally. The canal walls are lined with cilia which slow down the passage of the ovum and the sperm in order to increase their chances of meeting. On the other hand, the cilia also increase the stagnation of fluids and promote tubal inflammations. Situated in the upper portion of the broad ligament between the ovary and the round ligament, each uterine tube is an extremely mobile structure. Its medial part comes in contact with the small intestine, bladder and rectum. The lateral part contacts the internal iliac vessels, ureter, small intestine or sigmoid colon and rectum. This explains why inflammation of the tube can spread to these organs (e.g., peritonitis, proctitis, etc.), and why adhesions of the tubes are common.

The medial end of the uterine tube is connected directly to the superolateral angle of the uterus via a narrow (1mm) uterine ostium. The expanded lateral funnel-shaped end of the tube (the infundibulum) opens into the peritoneal cavity near (but not touching) the ovary. The opening, called the abdominal ostium, is 2-3mm in diameter and surrounded by many long processes called fimbriae, which are lined with a mucous membrane. This is a relatively unique situation — one mucous cavity connecting with another. Because the abdominal ostium is not in actual contact with the ovary, ova may occasionally pass into the peritoneal cavity rather than the uterine tubes, or the microbes of the uterine cavity and the uterine tubes may enter the peritoneal cavity. Occasionally, sperm may fertilize the ovum directly in the ovary (an ovarian pregnancy).

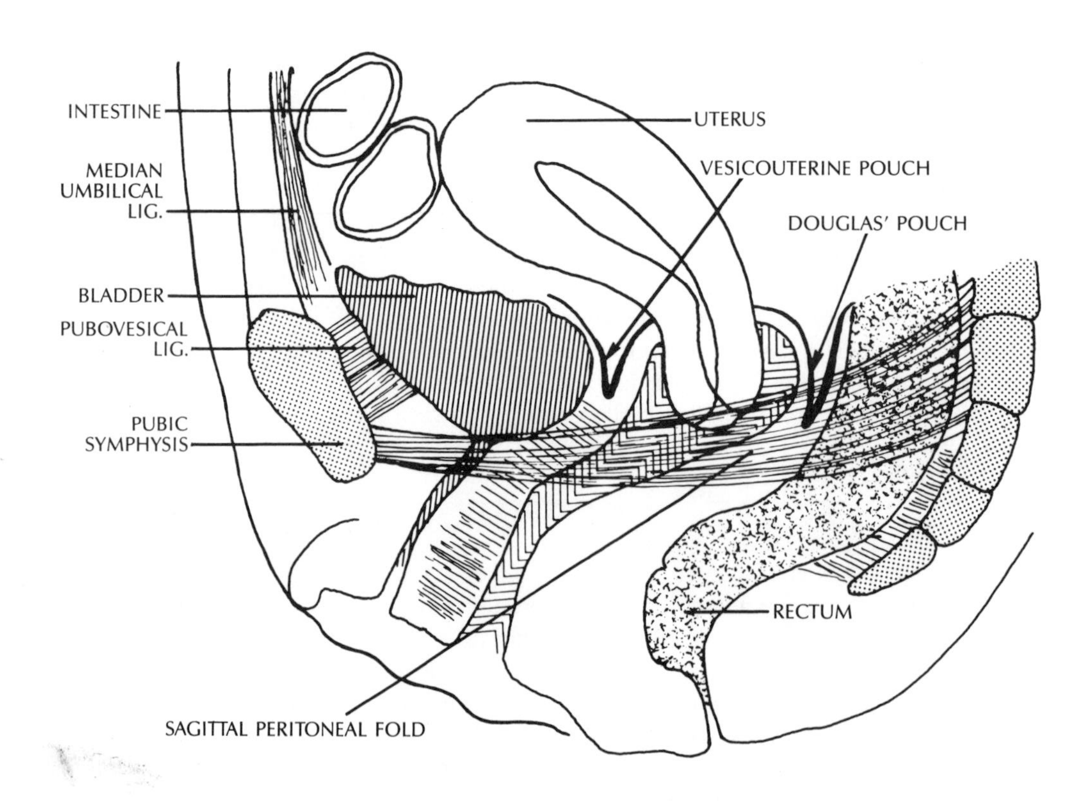

Illustration 10-1
Anatomical Relationships of the Uterus — Sagittal View

The ovaries are quite small — approximately the size of an almond (3.5cm long, 2cm wide, 1cm thick), and weighing around 8g. Each ovary is located in the retrouterine cavity posterior to the broad ligament, posterosuperior to the uterine tube and anterior to the rectum. It adheres to the posterior side of the broad ligament by means of a short peritoneal fold called the posterior fin which enables it to move vertically. It is connected to the uterine cornu by the ovarian ligament, to the uterine tube by the tubo-ovarian ligament and to the lateral wall of the lesser pelvis and the lumbar aponeuroses by the suspensory ligament. These ligaments, except for the latter, have a minor role in keeping the ovary in place. The stability of the ovary depends in large part on the uterus which, as we have seen, is very mobile. In multiparous women, the ovary often falls in the retrouterine cavity and can even fall into Douglas' pouch.

RELATIONSHIPS

One can distinguish two aspects, three edges and three angles of the uterus *(Illustrations 10-1 and 10-2).* The anteroinferior aspect is slightly convex and covered by the peritoneum which goes down as far as the isthmus. By folding back onto the bladder, it forms the vesicouterine cul-de-sac. In this way, the uterus leans on the bladder. The posterosuperior aspect is more convex, divided by a crest and is entirely covered by the

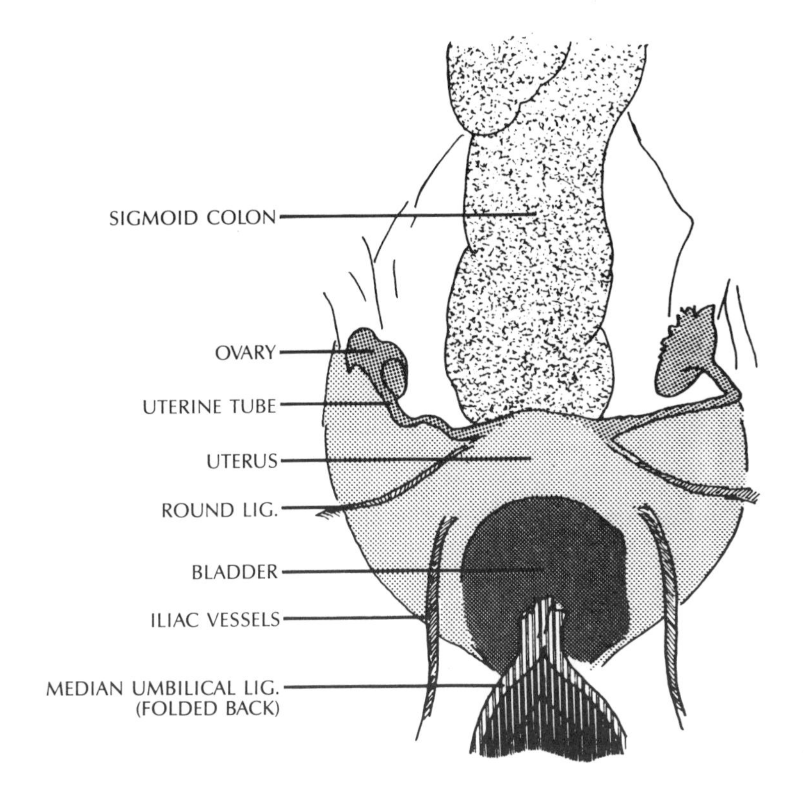

Illustration 10-2
Anatomical Relationships of the Uterus — Anterior View

peritoneum, which goes down as far as the upper part of the vagina. Douglas' pouch is formed as the peritoneum folds back from here onto the rectum.

The lateral edges are wide, rounded and press on the broad ligaments. The upper edge, which is rounded and coated by the peritoneum, is in contact with the small intestine and/or the sigmoid colon.

The lateral angles continue with the isthmus of the uterine tube. The round and ovarian ligaments leave from here. The isthmus interacts with the upper edge of the bladder and the vesicouterine cul-de-sac anteriorly. The supravaginal segment of the cervix, found in the pelvic subperitoneal space, is considered the third angle. Anteriorly it is in contact with the bladder, posteriorly with Douglas' pouch and thereby with the rectum, and laterally with its arteriovenous supply and the ureters. The intravaginal segment of the cervix enters the superior vagina and is pierced by the external opening of the uterus, called the external os.

VISCERAL ARTICULATIONS

The motion of the uterus is limited by the peritoneum, ligaments, vessels and perineum — that is, by the same tissues that suspend and support it.

Suspension

The peritoneum which covers the posterior bladder goes on to cover the uterus. It is most adherent at the level of the fundus (superior end), which is of importance in certain deviations of the uterus caused by adhesions with neighboring organs. The fact that the fundus is the part of the uterus most accessible by an abdominal route lends support to the possible efficacy of transabdominal manipulations.

The broad ligaments join the uterus to the pelvic floor. They are mobile in their connection to the uterus and follow the normal uterine anteversion. The posterior sides have close contact with the intestinal loops. This ligament can be divided into two levels. The superior level has three peritoneal fins created by the passage of the serous membrane on the round ligament, ovarian ligament and uterine tube. The inferior level of the broad ligament is stronger and is sometimes called the cardinal ligament. At this level, the anterior and posterior peritoneal layers move apart to accommodate a thick cellular layer called the parametrium. It consists of adipose tissue with some connective tissue and fibromuscular elements, and contains the ureter, uterine artery and uterine veins.

The round ligaments are long, thin fibromuscular cords stretched from the uterine cornua to the labia majora. They run in a sagittal plane and assist in orientation of the uterus.

The ovaries, which are not covered by visceral peritoneum, move freely in the pelvic cavity. They receive limited support from the ligaments described on page 238.

Support

Subperitoneal tissue concentrated around the vessels in this area gives rise to fibromuscular structures which help support the pelvic organs.

The sagittal peritoneal folds are aponeuroses disposed in a sagittal plane from the sacral foramina to the pubic bones. They adhere to the rectum, cervix, vaginal dome and base of the bladder, and interconnect these organs closely. Anteriorly, they are embedded in the subperitoneal tissue mentioned above. Posteriorly, from the superior third of the cervix to the anterior sacral foramina, they uplift Douglas' pouch to create two fibromuscular folds (called the uterosacral ligaments) which run from the uterus to the sacrum (where they insert around the third and fourth foramina) via the rectum.

These uterosacral ligaments are 6-7cm in height, and help form the external wall of Douglas' pouch. They are deeper and thicker superiorly where there are more muscular fibers; the inferior portion is mostly collagenous. They constitute the most solid part of the sagittal peritoneal folds, and oppose the displacement of the cervix toward the pubic symphysis. When the uterus moves, the bladder and sacrum tend to move with it because of these ligaments.

The cervix, to which the peritoneal folds are attached, is the fixed point of the uterus: version and rotational movements take place around it. For example, in retroversion, the entire uterus swings posteriorly around the superior part of the cervix such that the fundus points posteriorly and the cervix points towards the pubic bones (see also page 243).

There are three systems (superior, inferior and middle) of uterine support involving the structures described above. The uterus is more supported than suspended. The round ligament has a suspensory function only after ligamentopexy *(Contamin et al. 1977)*. The uterus is held up at the level of the cervix where the broad ligaments and sagittal peritoneal folds cross. One can therefore subdivide the superior support system into transverse and longitudinal elements.

The transverse elements of the superior support system are derived from the broad ligament and related fasciae. The fins and parametrium of the broad ligament were mentioned above. The posterior fin consists of the ovarian and tubo-ovarian ligaments, which are indispensable for good tubo-ovarian mobility. The parametrium, which plays an important role in anchoring the pelvic wall to the lateral edges of the uterus, has two distinguishable parts: (1) the internal fibromuscular part, which is attached to the supravaginal cervix, includes the tendinous center where the uterosacral ligaments are inserted posteriorly, a column around the cervix medially, and an external vesical column anteriorly. Numerous veins run across this part of the parametrium; the interaction between muscles and veins plays an important role in return circulation. (2) The external connective tissue part helps maintain the vesicovaginal veins in a horizontal position. Other transverse elements are the precervical fascia, which runs from the cervix to the anterior vaginal fornix and is continuous with fascia between the bladder and vagina; and the retrocervical fascia, which relates to the uterosacral ligaments, some fibers of which cross to the opposite side and form a uterine torus at the level of the isthmus.

The longitudinal elements of the superior support system consist mainly of the sagittal peritoneal folds. The vesicouterine ligaments are fibromuscular structures which help to maintain the parametrium by linking uterosacral fibers. They have no uterine insertion but go from the tendinous center of the broad ligament to the vesical cornu, where they are continuous with the pubovesical ligaments.

The inferior support system for the uterus consists of the levator ani and perineal muscles upon which the cervix, via the vagina, rests. The primary supportive element of the perineum is the fibrous nucleus.

The middle support system, which joins the superior and inferior systems, consists of the fascia between the bladder and vagina, and the sagittal peritoneal folds, sagittal sheaths of fascia connecting the bladder, uterus and rectum. The fibromuscular architecture of the pelvis results in several weak points here, notably the posterior vaginal groove. This is due to separation of the uterosacral ligaments and insufficient aponeurosis in the sagittal peritoneal folds and between the vagina and rectum.

Sliding Surfaces

Via the peritoneum, the uterus articulates with the bladder anteriorly and the rectum posteriorly. As mentioned in chapter 9, neither the bladder nor the uterus can be manipulated without affecting the other. As discussed under "Manipulations" below, we manipulate the uterine suspensory system mostly via the bladder. The uterus also articulates with the small intestine and sigmoid colon. Loops of the small intestine sometimes infiltrate the vesicouterine cul-de-sac. Ptosis of the organs above the uterus may compress it, accentuating the version phenomena which are discussed on the next page.

TOPOGRAPHICAL ANATOMY

The inferior aspect of the uterus is usually located 2-3cm above the plane of the pelvic outlet. A transverse plane that goes through the superior aspect of the pubic symphysis always meets the uterus, normally at the level of its inferior third, sometimes even lower. It is therefore easy to explore the bottom of the uterus by pressing in the abdominal wall just above the superior edge of the symphysis.

The supravaginal segment of the cervix is in contact with the bladder, and posteriorly with the rectum (via Douglas' pouch) at the level of the sacrococcygeal articulation. The lateral projection of the cervix falls slightly in front of the ischial tuberosity.

The ovary is usually found on a line joining the anterior superior iliac spine (ASIS) to the superior edge of the pubic symphysis, just medial and slightly superior to the medial edge of the psoas muscle.

Physiologic Motion

UTERINE MOTION

Motion of the uterus in response to diaphragmatic respiration, though slight, does exist and may be seen during hysterosalpingography. The uterus is extremely mobile; gynecologists know only too well that from one examination to the next, the uterus changes its position depending on the menstrual cycle and the distention of the bladder and rectum. Sometimes a certain position is maintained and causes congestive and mechanical problems. We shall briefly describe some abnormal variations of the position of the uterus below.

The uterus makes a 60 degree angle with the umbilicococcygeal axis. Normally, when the bladder is empty, the entire uterus inclines forward (anteversion). Also, the longitudinal axis of the body presents as a curved line which is concave anteriorly (anteflexion), with an angle of 120-130 degrees between the body and cervix. The longitudinal uterine and vaginal axes should therefore both be concave anteriorly, and parallel to the sacral curvature. The position of the uterus is defined in relation to the midsagittal plane. The body and the fundus rest on the superior face of the bladder; any variation from this position will cause abdominal pressure to affect the uterus adversely. As the column of pressure becomes displaced, it presses on the perineum and causes a gradual separation of muscular and connective fibers, as well as local congestion.

TUBO-OVARIAN MOTION

We would like to briefly discuss the capture and transport of the ovum in order to emphasize the importance of good mobility, and lack of restrictions, for the proper functioning of the reproductive organs. Bear in mind that the tubes are so mobile that they can cross the midline to get an ovum from the contralateral ovary.

When ovulation begins, the tubal fimbriae go into a rhythmic movement. The contractile fibers of the suspensory and ovarian ligaments mobilize the ovary, so that it turns on its longitudinal axis, and lower or raise it depending on its position in the pelvic cavity. During these motions, the fimbriae reach the place where the ripe follicle will leave the ovary. The ovum enters and travels through the uterine tube under the influence of the fimbriae, contractions of the tube and the current produced by the cilia. Both ends of the tube must be clear for this current to be maintained. Imagine the consequences of an adhesion or other restriction anywhere in the vicinity of the ovary or uterine tube, and you will easily understand the importance of mobility in female fertility.

DEVIATIONS

There are many possible deviations of uterine position. We will mention only a few common types here. For more information, please consult standard reference works on obstetrics and gynecology.

Problems with the relationship of the uterine body to the cervix result in what we call intrinsic deviations. With too much anteflexion, the body is too far forward and its angle with the cervix is therefore too acute. The other common deviations of the uterus are retroversion (a backward swing of the entire uterus) and retroflexion (a posterior curvature of the longitudinal axis of the body in relation to the cervix). The most common condition in women over the age of 50 is retroversion combined with retroflexion, in which both the cervix and the uterus swing backward, and the cervix lies on the rectum. At the same time, the small intestine lies on the bladder, pushing it down.

In lateroflexion, the cervix is bent to the side (most commonly the left side). This phenomenon is relatively rare and of little interest to gynecologists, but of great interest to us as a subject for manipulation. Lateroflexion is usually due to lateral fixations of the uterosacral ligaments or problems of the cecum or sigmoid colon.

The extrinsic deviations involve displacement of the entire uterus and its surrounding tissues within the pelvic cavity, e.g., far forward against the pubis. The most common and important extrinsic deviation is uterine prolapse against or through the pelvic floor. Gravity is often the culprit, alone or in combination with delivery accomplished by forceps, suction or an overlarge episiotomy. In such cases, the pelvic fibromuscular system loses much of its elasticity and contractility, and has trouble recovering them. Some women appear to be predisposed to this problem. As the uterus becomes more vertical, it is more likely to move into the vagina. There are three stages of prolapse, the most severe of which may require surgical intervention.

MOBILITY

The uterus is a very mobile organ, with most of its motion taking place in the sagittal plane. This mobility is well demonstrated by the variety of positions it assumes under the influence of surrounding organs:

- if the bladder is full, the uterus is pushed posteriorly
- if the rectum is full, the uterus is pushed anteriorly
- if both are full, the uterus moves superiorly
- gravity tends to pull it downward
- rarely, fibrosis of the supporting structures can lead to lateral displacements

MOTILITY

The motion of the uterus is the same as that of the bladder. In expir, the uterus moves posteriorly and then superiorly, as it did during embryogenesis.

Indications for Visceral Manipulation

In our opinion, every patient who undergoes an operation on the urogenital system should consult an osteopath postoperatively after three menstrual cycles. We refer to surgery for cysts, fibromas and neoplasms, as well as ligamentopexies, caesarean sections and even colposcopies. The latter procedure is often ignored because it seems as harmless as hysterosalpingography or ultrasound. However, it is an invasive procedure with subsequent irritation and possible scarring from the healing process. Anytime a serous membrane is opened, it tends to become irritated and undergo adhesions and other restrictions, which then disturb the mobility and motility of the organs it contains.

The risk of mechanical trauma to the reproductive system is high following obstetrical procedures, and depends to a great degree on the gentleness of the obstetrician or midwife. Postpartum cystoceles are also very common. Teach your patients that suction and large episiotomies are neither necessary nor desirable for safe delivery and post partum recovery!

The female reproductive system is constantly under attack by microorganisms such as bacteria, viruses and yeasts, which often produce reproductive dysfunction (dysmenorrhea, infertility, etc.) secondary to tissue damage.

Problems of pelvic stasis are caused by phenomena such as old age, hypotonia, weight loss, hormonal disorders and sedentary life style. Presenting symptoms involve pelvic circulation problems provoking discomfort in the lower abdomen, and are often linked to menstrual pain, hemorrhoids or varicosities of the veins of the lower limbs. Increased intrapelvic pressure will trigger the symptomatology, and pressing on the pelvis will cause pain. The patient often has leukorrhea, polyuria, cystitis and lower back pain. Factors such as constipation, overeating or tight clothing aggravate these symptoms, and elevation of the abdomen relieves them. One example of pelvic stasis is when the uterus collapses posteriorly and the shifting of the weight of the adnexa and fundus make the uterus swing toward Douglas' pouch. This causes inflammation with risks of adhesions to the peritoneal membrane. We believe that lower back problems (particularly in women) are usually secondary to the problems of urogenital organ position rather than to the sacrosanct disc disorders. When dyspareunia is not primary but occurs postpartum or after a fall, it is usually due either to fluid congestion and vasomotor problems, or trauma to the sacrococcygeal area.

Evaluation

Again, the history is of primary importance and should enable you to differentiate important functional syndromes from structural problems. It is essential, when there is any doubt, to consult a gynecologist and perform any necessary test or examination. Perhaps in this area, more than any other, you must be able to recognize your own limitations.

The bimanual pelvic exam allows you to appreciate various uterine positions, differences in endometrial elasticity, the possibility of pregnancy, etc. You should be able to easily and painlessly mobilize the uterus during palpation.

The intrarectal exam enables you to feel the position of the cervix, the pressure states of the different cul-de-sacs and the tension of the cervicosacral attachments. We shall discuss this exam more fully, in relation to the coccyx, in chapter 11.

Laboratory examinations are indispensable if there is any doubt as to the etiology of symptoms. These include simple ultrasound as well as colposcopy. Most women are

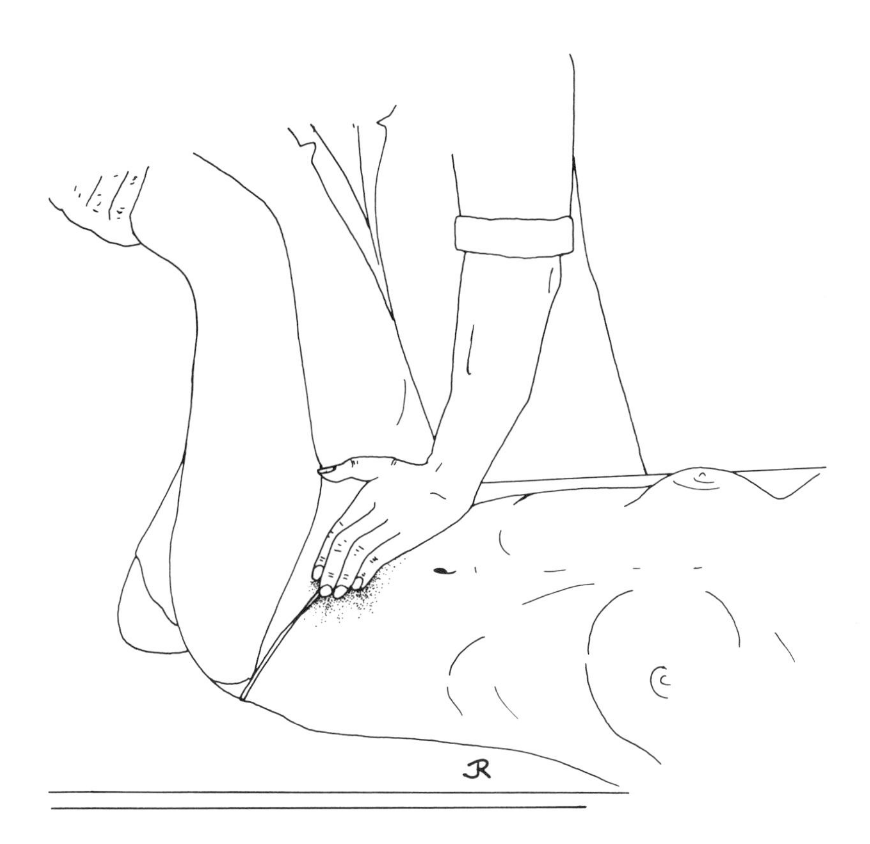

Illustration 10-3
Mobility Test of the Uterus — Abdominal Route

examined regularly by a gynecologist, and undergo an annual pap smear, which considerably limits the risks of a late discovery of neoplasia. You should always be aware of the considerable risk of neoplasia in the urogenital region.

MOBILITY TESTS

Abdominal palpation is performed with the patient in the supine position, legs bent and feet on a cushion to increase flexion of the hip. The aim is to approach the fundic region laterally in order to check the mobility of the uterus. Place the fingers just above the pubic symphysis on the inferior origin of the rectus abdominis muscles in order to direct them posteriorly on the bladder. The more the legs are bent, the more deeply you can go into the pelvic cavity. You can even actively bend the legs to increase the efficiency of your palpation *(Illustration 10-3)*. To reach the uterus, you need to go via the abdominal wall, the loops of the small intestine, or the bladder, depending on their position in the body.

The mobility of the ovary and broad ligament is tested in either the lateral decubitus or reverse Trendelenburg position with the hips and knees flexed. Starting at the ASIS, push your hand posteriorly and, when you have gone as far back as possible, move it toward the symphysis. Evaluate the elasticity of the tissues and note zones of adhesion,

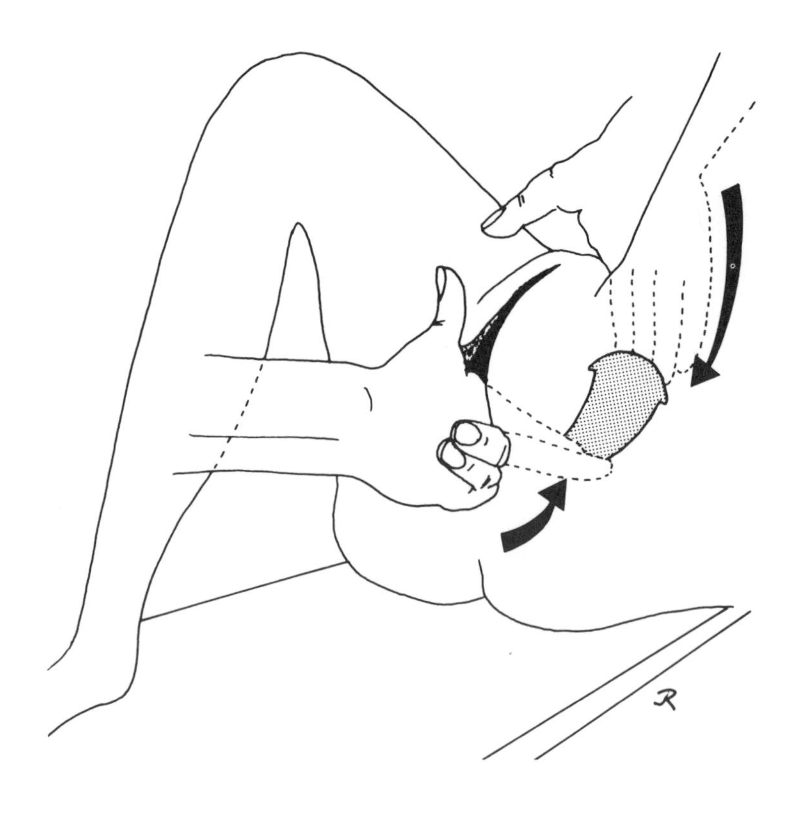

Illustration 10-4
Mobility Test of the Uterus — Bimanual Palpation

always comparing one side to the other. If you lose your place, ask the patient to contract the psoas muscle, which can serve as a reference.

For bimanual palpation, place two fingers into the vagina and the other hand on the abdomen facing the fundus. The two hands act together to test the mobility of the uterus and ovaries *(Illustration 10-4)*. In the case of a serious retroversion, the abdominal hand will not be of much help, as the fundus will not be palpable.

Rectal palpation is performed in the prone position. When your finger reaches the level of the sacrococcygeal articulation, push it forward and, via Douglas' pouch, it will contact the uterus. If the uterus is retroverted, you will feel a round hard mass which prevents your finger from moving forward. When in a normal position, the cervix shows little resistance to digital pressure and is easily movable.

The sacral pressure technique provides an easy method for distinguishing musculoskeletal sacral and sacroiliac restrictions from those resulting from tightness of the uterosacral ligaments. With the patient prone, push the sacrum anteriorly with the heel of your hand at the S2/3 region. If there is no movement, there is a musculoskeletal restriction. To be more precise, test the three (superior, middle and inferior) areas of the sacrum. If the sacrum moves anteriorly but has difficulty in returning posteriorly or does so slowly, there is a restriction in the uterosacral ligaments. When the return motion has a rotatory nature, there is a restriction on one side (usually the side toward which it rotates).

Normally, the uterus leans on the bladder. Vesicouterine mobility can be tested by lifting the bladder via the median and medial umbilical ligaments. One cannot separate the bladder from the uterus, but a force applied to one automatically has an effect upon the other. The patient can be lying down or seated; we prefer the latter position because the test is facilitated by gravity (see pages 226-28).

MOTILITY TESTS

The test for the uterus is similar to that for the bladder. The patient is supine with knees bent. Place the palm of one hand over the pubic symphysis with the fingers pointed toward the umbilicus. During expir, the back of the hand moves toward the umbilicus (the fingers move posteriorly while the heel of the hand moves anteriorly) *(Illustration 10-5)*. For a fuller appreciation of uterine motility, place your other hand under the sacrum with the fingers under the base and the heel of the hand under the apex. In expir, that hand should move anterosuperiorly (the base of the sacrum moves posteriorly and the apex anteriorly).

The motility test for the ovaries is performed with the patient in a supine position. Place the palm of the hand between the ASIS and the symphysis, fingers directed superiorly and slightly laterally. During inspir, the hand will rotate laterally and move slightly superiorly (a clockwise rotation on the left, counterclockwise on the right). Once you are able to feel this motion comfortably, you will also be able to feel the lateral edge of the hand move posteriorly, with a small component of external rotation. The opposite motion occurs during expir.

Restrictions

Possible restrictions range from a partial absence of mobility to a total collapse of the reproductive system, and we cannot describe them all. The reproductive system

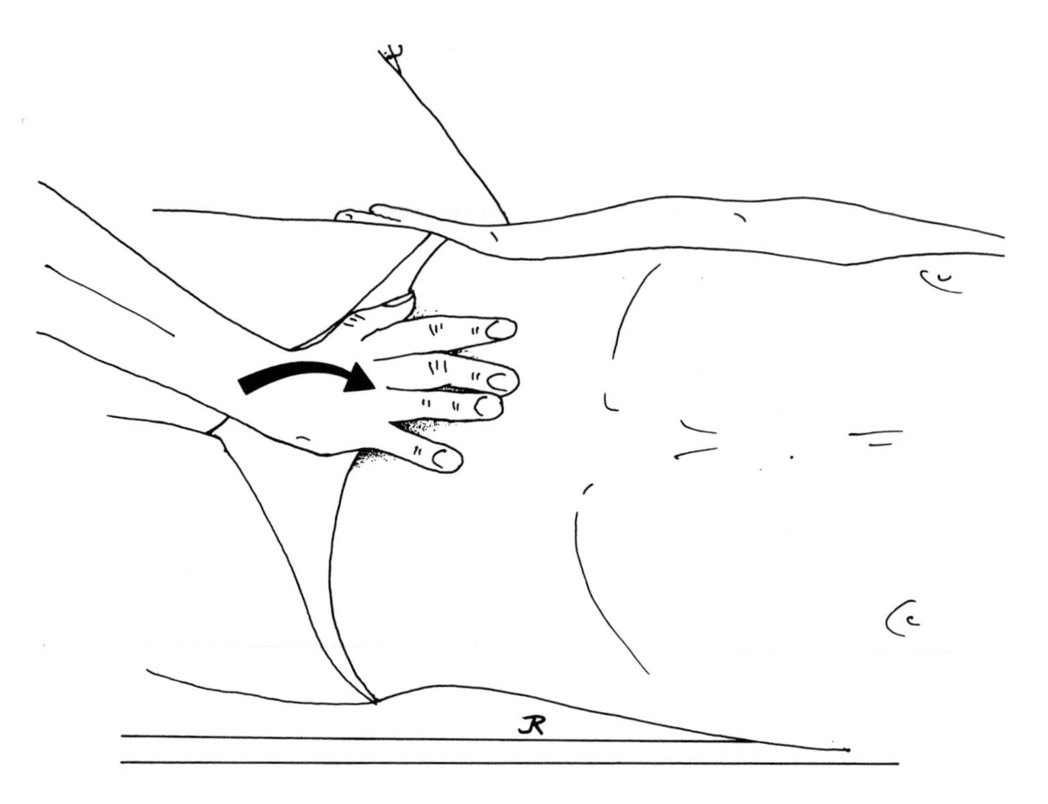

Illustration 10-5
Motility Test of the Uterus

is dynamic; it is not held together by simple ligaments stretched out like pieces of string. Nearly the entire system of attachment is based on contractile fibers; the broad ligaments, round ligaments, parametrium, subperitoneal tissue, sagittal peritoneal folds and pelvic floor all contain some contractile muscle fibers. For this reason, induction is most effective on the reproductive system — it is much easier to induce a release in a muscle than a ligament. It seems that, on a pelvic level, problems with motility involve the tonicity of the contractile fibers more than their mobility.

Scars from abscesses or surgery, as well as microadhesions from infectious processes, disrupt the motion of the reproductive organs. Small twists or spasms can reduce or close the abdominal ostium and interfere with the complex phenomena of ovulation and fertilization. Miscarriages, abortions and ectopic pregnancies can all cause serious tubal adhesions. Reflex spasms are complex phenomena and depend on both local conditions and systemic states. For example, emotional upset during hysterosalpingography can prevent passage of the contrast material. This demonstrates that the fear of being infertile can interfere with the effort to become fertile.

Manipulations

In manipulations of the pelvic organs, a good general rule is to treat the abdominal organs first. For example, you should clear up any problems of the small and large intestines before beginning work on the uterus. Also, be sure that the patient urinates before treatment.

CONTRAINDICATIONS

A patient who has an IUD or is pregnant should not have genital manipulation under any circumstances. All infections and hemorrhagic symptoms necessitate much care and a precise diagnosis. Do not proceed if you have any doubts.

DIRECT AND COMBINED TECHNIQUES

The direct technique in the supine position is similar to the mobility test; the patient has her feet raised on a cushion in order to increase hip flexion, release muscular tension and increase the depth of the field to be palpated. Begin by pushing or pulling the fundus from side to side with a translatory movement to effect a local release. This

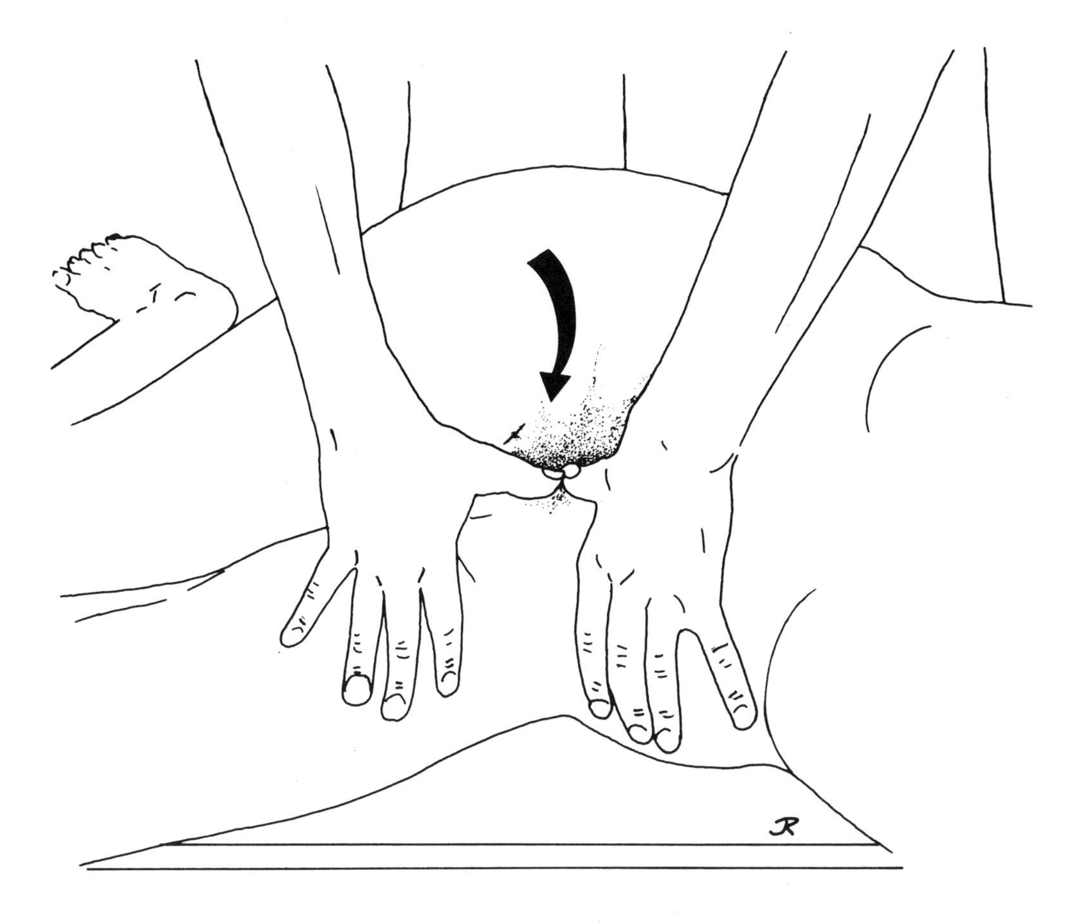

Illustration 10-6
Manipulation of the Uterus Via the Abdominal Route — Lateral Decubitus Position

is an efficient defibrosing technique, and should be repeated about ten times, gently and rhythmically without irritating the tissues, until a release is felt. If a release does not occur after 10-15 cycles, either the problem is someplace else or you are doing the technique incorrectly. In any case, stop, as further treatment will only irritate the local tissues. You can increase the efficiency of the stretching by sidebending the lower limbs on the pelvis. Fix a point (preferably a point of restriction) of the lateral vesicouterine region and focus on it more clearly by sidebending around it to the opposite site by using the lower limbs. This produces an excellent local stretch. This technique is especially effective on the broad ligaments and tubo-ovarian adhesions.

In the lateral decubitus position, the patient has the hips and legs flexed on the body to release the abdominopelvic wall. Depending on the restriction, either stretch the lateral aspect of the uterus closest to you by pushing down with your thumbs, or bring the opposite region toward you with your fingers *(Illustration 10-6)*. Repeat this gently and rhythmically until you feel a release.

In the reverse Trendelenburg position, use the technique described for the supine position, with the addition of an antigravitational component. That is, the hands mobilize the uterus superiorly as well as laterally *(Illustration 10-7)*. This is the most effective technique in cases of urogenital ptosis.

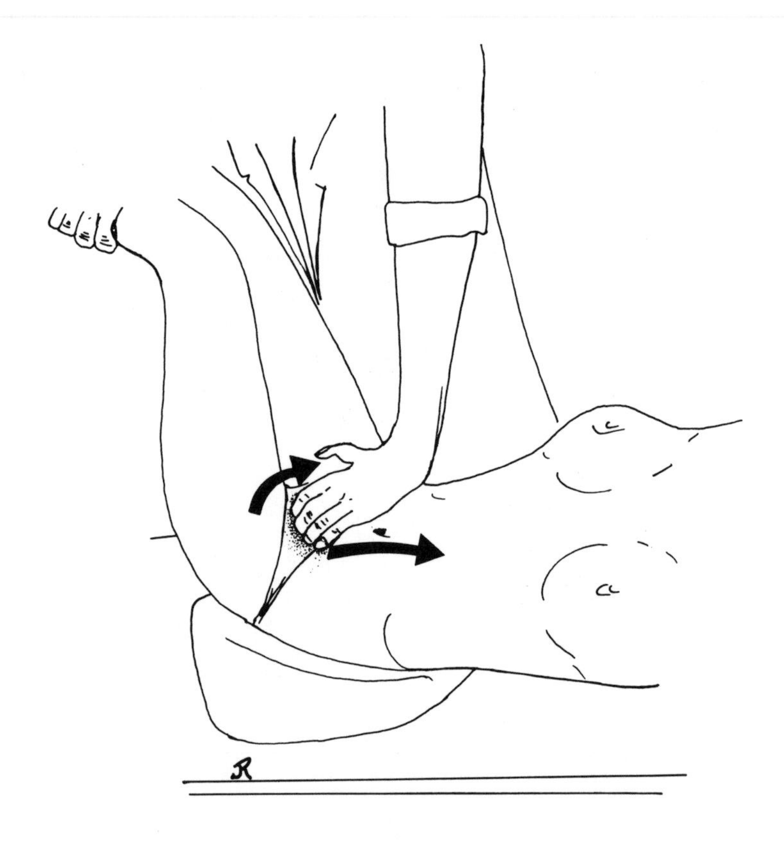

Illustration 10-7
Manipulation of the Uterus Via the Abdominal Route — Reverse Trendelenburg Position

In the seated position, the technique is similar to that for bladder restrictions. Place your fingers just above the symphysis to pull the median and medial umbilical ligaments upward. To increase the stretching effect and thus the efficacy of the technique, bend the patient backward to increase the distance between the xiphoid process and the symphysis *(Illustration 10-8)*. This technique releases pressure from loops of the small intestine which often contribute to ptosis of the pelvic organs. This is an example of the general rule to clear what is above before attacking problems in the pelvis.

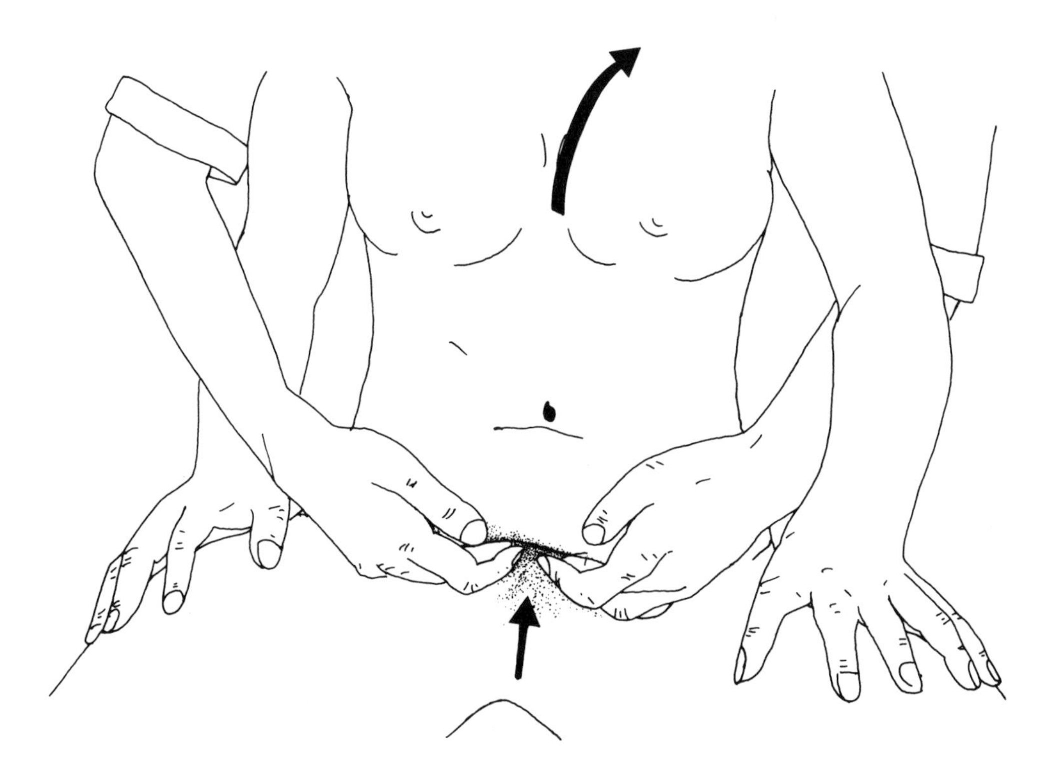

Illustration 10-8
Vesicouterine Manipulation — Seated Position

INDUCTION TECHNIQUES

For uterine induction with the patient in the supine position, place the palm of your hand just above the symphysis and then push it posteriorly as if you wished to go under the pubic bones. Gradually direct your hand upward. This technique can also be performed with one hand under the sacrum, which should move in the opposite direction (anteroinferiorly) to create opposing forces. This technique is intermediate between direct pressure and induction in terms of force, and enables overly tight uterosacral structures to relax. For lax structures, we do short and rapid stretching in an attempt to stimulate them. We have to admit, however, that while our methods can produce beneficial changes temporarily in cases of lax structures, it is difficult to maintain the changes.

With the lateral decubitus position, place one hand above the symphysis and the other on the sacrum. Carry out an anteroposterior induction/pressure in the same directions as in the preceding technique *(Illustration 10-9)*. The advantage of doing the technique in this position is that it frees the hand on the sacrum (in the supine position, the hand is squashed by the patient's weight), and enables you to easily combine movements in all three planes.

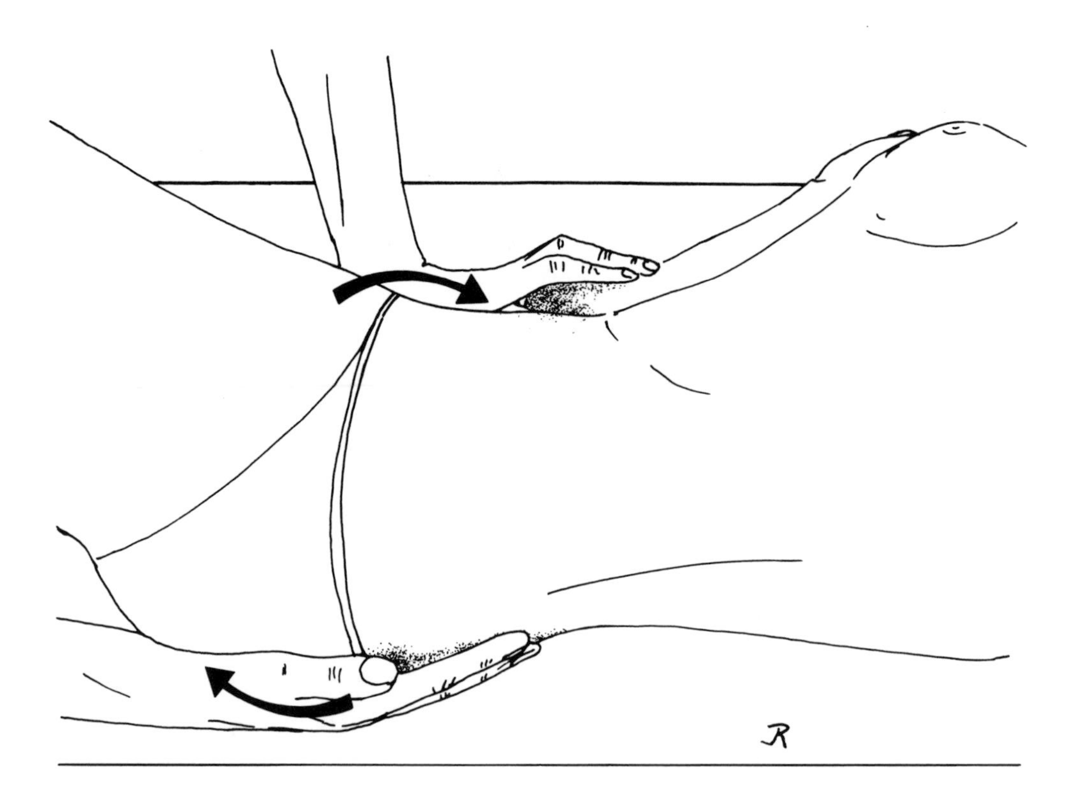

Illustration 10-9
Uterosacral Manipulation — Lateral Decubitus Position

With the vaginal bimanual technique (page 245), you can free uterine fibroses and, above all, stretch the cul-de-sacs. This technique should be performed gently and rhythmically, and a release should be felt after five or six cycles. In order not to confuse the doctor-patient relationship, this technique should not be needlessly prolonged.

Induction via the rectal route is definitely the most effective technique for manipulation of the sacral structures and the release of Douglas' pouch. The intrarectal finger should push the cervix anteriorly and make the sacrococcygeal unit move posteriorly. This technique is described in detail in chapter 11.

Manipulation of the uterine tubes and ovaries must be gentle. For this reason, we prefer using induction on them to avoid the risk of damaging the fragile structures involved. With the patient in the supine position with legs bent, place the palm of your hand on the line between the median axis and the left or right ASIS, depending on which side is being worked on. For the left ovary, the palm of the hand moves slightly superolaterally during inspir and then performs a clockwise rotation *(Illustration 10-10).* Manipulation of the ovaries must always be accompanied by manipulation of the uterus; the uterus is usually treated first as it is made up of stronger tissues.

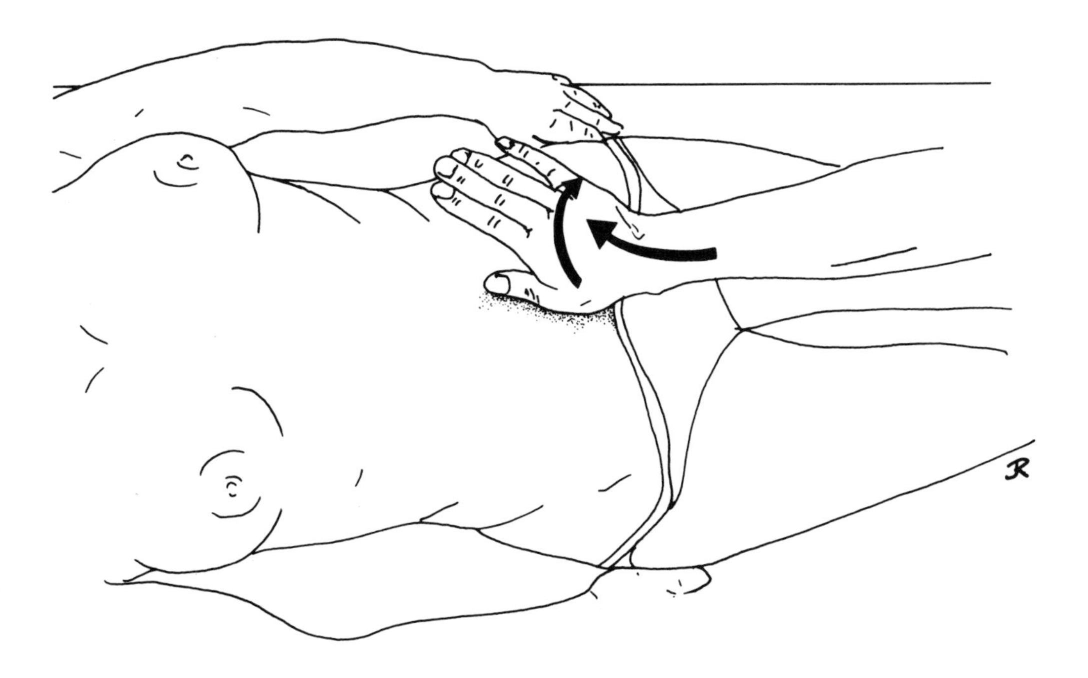

Illustration 10-10
Tubo-Ovarian Induction

Effects

All serous membranes tend to undergo adhesions after certain inflammatory problems. The female pelvic organs are very vulnerable to these inflammatory processes and tend to often ptose or congest. The effect of our techniques is to defibrose, decongest, raise and remobilize the reproductive organs. Whenever there are scars, restrictions of mobility or poor circulation, urogenital manipulation is useful. The more intense and chronic the suffering, the better the results. This type of case forms a major part of our own practices; one of us (Barral) has treated over five thousand women for gynecological problems.

The aim of the tubo-ovarian manipulations is to increase or restore local mobility; any idea of using a strong force must be abandoned. We are practicing visceral clock making — even a slight fold of the fimbriae, bend in the tube, bad orientation of the ostia or uterine sidebending can be enough to block pelvic fluid circulation and inhibit the process of fertilization.

We believe that we have resolved several cases of infertility through visceral manipulation, although this kind of result can never be strictly evaluated scientifically, as too many parameters are involved. Our best results have occurred with patients who came in for a different problem. In any case, it seems very reasonable to attempt to harmonize and release the soft tissues of the pelvis before performing complicated and expensive testing procedures.

After intravaginal manipulations, it is not uncommon for there to be a slight salmon-colored bleeding for a few days. This is a transient problem and will resolve without further intervention. Be sure to check the area posterior to the cervix, as many important and surprising things (old secretions, remnants of yeast infections, condoms, etc.) can be found here.

Adjunctive Considerations

ASSOCIATED OSSEOUS RESTRICTIONS

Lumbosacral vertebral restrictions are almost always found in patients with urogenital problems, either due to reflex or mechanical effects. Reflex knee pain may be provoked by an irritation of the genitocrural nerves from problems of motion inside the pelvis. Hormonal disorders are frequently responsible for reproductive problems. In this case, the vertebral restrictions are localized in the superior cervical area, perhaps because of their relationship with the hypothalamic-pituitary axis.

RECOMMENDATIONS

Be aware of the patient's menstrual cycle — the best time to use manipulation is at the beginning of the cycle just after the menses have ended. Remember that either an IUD or pregnancy is a strict contraindication to manipulation of the pelvis. Tampons sometimes decrease the mobility and motility of the cervix, especially with retroversion. They can be helpful for some types of incontinence. For problems of ptosis and anteversion, we recommend that the patient use the reverse Trendelenburg position as described in chapter 5 (page 128). Application of heat to the abdominopelvic region may also be useful. Advise your patients to empty the bladder often; a full bladder pushes the uterus posteriorly and strains the surrounding tissues.

Chapter Eleven: The Coccyx

Table of Contents

CHAPTER ELEVEN

The Coccyx

For those of you who have not read the preceding chapters, it may seem strange to find a chapter on the coccyx in a book about viscera. We agree that the coccyx is not a viscera; however, we have found that problems of sacrococcygeal articulation often have visceral repercussions. Because of the excellent results we have had with manipulation of the coccyx in relation to visceral manipulation, we think that our readers will find this chapter very useful.

Relationships to Viscera

The sacrum and coccyx connect anteriorly with the rectum, the prostate in men and uterus (via Douglas' pouch) in women and, further forward, the bladder. These relationships are important in the treatment of many urogenital disorders.

The sacrococcygeal articulation is a diarthrosis with an articular motion of up to 30 degrees. It is surrounded by the anterior, posterior and lateral sacrococcygeal ligaments. In the normal subject, these ligaments allow good mobility while maintaining appropriate tension on the coccyx and associated structures.

The coccygeal ligament is very important. It is the only manipulable ligament which enables one to have a direct effect on the dural tube. (The dura mater is also attached to C2; osteopaths are well aware of this peculiarity.) This partly explains the repercussions of mechanical perturbations of the sacrococcygeal articulation on the primary respiratory motion.

Almost all the soft tissues of the pelvis are attached to the coccyx. The connections are listed below to make you aware of the possible distant repercussions of a sacrococcygeal articular lesion:

- anococcygeal ligament
- sacrotuberous ligaments
- sacrococcygeal ligaments
- coccygeus muscle
- levator ani muscle
- some gluteal fibers
- some fibers of the prerectal hemisheaths

Physiologic Motion

The sacrococcygeal articulation is sometimes classified as an amphiarthrosis rather than a diarthrosis. We disagree. Having tested hundreds of sacrococcygeal articulations by rectal manipulation, we affirm that it is a mobile, even *very* mobile joint. It has an amplitude of 30 degrees; any limitation signifies a restriction. Its best-known movement takes place during the birth process, when it plays an important role in expansion of the pelvic outlet. Posterior flexion of the coccyx is involved in the relaxation of the levator ani muscle and the pushing of the child's head. A functional sacrococcygeal articulation allows some vital centimeters to be gained at this crucial time.

The sacrococcygeal articulation also has a physiological role in copulation, defecation and micturition. It plays an integral part in lumbosacral dynamics; problems with the coccyx can contribute to lumbosacral restrictions.

DISPLACEMENTS AND RESTRICTIONS

The sacrococcygeal articulation is one of the few where positional lesions occur. We have stated *(Barral et al. 1981)* our aversion to osteoarticular terminologies which describe lesions with words such as posterior, displacement, subluxation, etc. Our concept is that an osteopathic lesion (also known as somatic dysfunction) is a lesion of restricted motion. The coccyx is an exception to this general rule. Palpation and radiography leave no doubt that there is displacement at this joint. In our experience, in 80 percent of patients with this problem the displacement is anterior, which automatically lessens the sagittal diameter of the pelvic outlet. The sacrococcygeal articulation can be restricted to varying degrees following fibrotic retractions of the soft structures which either maintain it or attach themselves to it.

A seemingly harmless coccygeal displacement can have serious consequences upon pelvic physiology. In chapter 9, we mentioned Huguier and Bethoux's (1965) experiments on the displacement of the urethrovesical junction and its physiological consequences. When the coccyx moves closer to the pubic symphysis, the musculoaponeurotic fibers from the pubis to the coccyx become so slack that they lose their tonus. If the origin and insertion of a muscle move closer together, a great portion of the muscle's power is lost.

When the bladder and rectum are relaxed in this way, incontinence may result. One of our patients became incontinent of stool (and impotent) after a fracture of the coccyx. Treatment of the coccyx resulted in a cure of both of these problems.

Reproductive function can also be affected by this mechanism. Examples of problems induced include unsatisfactory sex and weak or transient erections. The perineum is very important in genital vasomotor responses; its disturbance can cause congestive problems. Take the example of uterovaginal mobility in orgasm. At this time, the shape of the vagina is modified, with two thirds of the internal fibers lengthening while the exterior third of the wall contracts and reduces its diameter. At the same time, the uterus moves up in the pelvic basin and becomes vertical. All these pelvic structures must be mobile in order for this process to take place. It is not uncommon for a woman to experience decreased libido following a fall on the coccyx or a laborious childbirth; a simple sacrococcygeal manipulation may resolve this problem.

To summarize: with anterior fixations of the coccyx the fibers of the pelvis shorten, easily leading to various types of prolapses. All other coccygeal fixations cause a general tightness of the pelvic ligaments.

Indications for Evaluation

Are urogenital problems which begin after a fall onto the buttocks, a car accident or difficult labor usually due to coccygeal problems? The relationship between cause and effect in these cases is not always easy to establish. A fall on the coccyx could have happened ten years before. There is, however, a sign that is almost pathognomonic for coccygeal problems — the patient cannot stay sitting (in the car, at the theater, etc.) for any appreciable length of time. As one sits down, the ischia move apart to enlarge the ischial base. If the coccyx is restricted, the sacrotuberous ligaments cannot stretch and the iliolumbar ligaments will be stretched when they should be relaxed.

Typical symptoms of a sacrococcygeal lesion in a female subject are inability to sit for long periods of time, declining quality of sexual relationships and cystitis. With inexplicable bladder infections, be sure to check coccygeal mobility. We have successfully treated many urinary tract infections this way. Urogenital ptoses, urinary incontinence and uterine retroversions are also important indications to check the coccyx. These dysfunctions do not disappear after manipulation, but do improve. In the male, prostatitis and hemorrhoids are often linked to a problems of the coccyx.

Fixations of the coccyx can have wide ramifications. Besides the urogenital dysfunctions outlined previously, the coccyx has a strong connection to other organs, particularly the kidneys. Also, a fixation of the coccyx can lead to a general decrease in the motility of the entire body, and it should be checked in people who are devitalized or suffering from general depression.

Evaluation

In evaluating lumbosacral and pelvic dysfunctions, it is essential that you free yourself from a limited mechanistic structural concept of vertebral restrictions (e.g., "low back pain equals a restriction of L5"). When a patient presents with low back pain, it is not sufficient to define the quality of the pain when taking out the garbage in the morning! Do not hesitate to ask about possible urogenital syndromes. Most patients have trouble talking about problems such as incontinence. If you are natural, at ease and, above all, do not let the patient feel your own excessive modesty, a verbal relationship can be formed without embarrassment or ambiguity. We will not offer you a standard questionnaire for gynecological examination — refer to your texts — but do, systematically, go deeper and in detail with your history.

MOBILITY TESTS

Test the mobility of the coccyx first via the following external route and use the rectal route only if necessary. The patient sits with the legs hanging slightly apart so that the ischiococcygeal attachments are taut. With your index finger, follow the gluteal fold to a point approximately 1cm posterior to the anus. Sidebending often helps in positioning the finger anteriorly. Bring the patient back to the neutral position. At the inferior extremity of the coccyx, transmit a slight posterosuperior force. If the sacrococcygeal articulation is injured, the patient will feel a sharp pain, which is immediate and excruciating, and which leaves no doubt as to the diagnosis. It is surprising to see that such a small movement can cause such intense pain. Carry out a rectal examination only if this test is positive, in order to fine tune the diagnosis. Less than 10 percent of subjects with a negative test (no pain) will have a coccygeal restriction.

For the rectal route, we do not like the embarrassing quadrupedal position which makes patients feel they are part of a show. Place the patient in a prone or lateral decubitus position with the legs slightly separated. Put on a lubricated glove and separate the buttocks with the free hand. When the index finger is on the anus, press gently without the force that would cause immediate penetration, and enter slowly. Go past the anal sphincter and leave the finger in the rectal ampulla with the pad posteriorly against the coccyx and sacrum. If there are external hemorrhoids, push them in at this point. Once the index finger is in the rectum, place the thumb parallel to it externally and move the coccyx, first anteriorly and then posteriorly. You will be surprised by the mobility of the sacrococcygeal articulation when it is free.

This test will enable you to determine if the coccyx is fixed (partial or total absence of movement), or displaced anteriorly (the coccyx is very anterior and painful when moved). Anterior restrictions are most common but lateral restrictions or displacements also occur. They are found by placing the finger on one side of the coccyx and pressing it toward the opposite side. This test enables you to evaluate the elasticity of the sacrotuberous and inferior sacroiliac ligaments.

Another way to test the sacrococcygeal and sacrotuberous ligaments is to push the sacrum anteriorly with the external hand. First evaluate the sacrococcygeus (by pushing the internal finger posteriorly) and then the sacrotuberous (by palpating the lateral aspects of the coccyx with the internal finger). There should be no restrictions in movement, and only slight tenderness.

Do not lose this opportunity to check the uterus or prostate. The cervix should be firm, but supple and movable. If it is posterior and cannot be moved anteriorly, there is a restriction of the uterosacral ligaments. Lateral motion of the cervix should also be evaluated. The same is true for the prostate. If you feel any hard, button-like areas over these tissues or in the rectum, refer the patient to be checked for the presence of cancer. Over 50 percent of all rectosigmoid cancers are within the reach of your fingers.

Certain rules and precautions must be observed in relation to a rectal examination. A few of the more important are:

- parents should be present during examination of children
- avoid rectal examinations during the menstrual period
- be sure to give clear explanations of your procedure
- do not attempt to force reticent patients; usually they are reticent because they do not feel clean, but will gladly cooperate at the following session

Manipulations

Sacrococcygeal manipulation is performed with one hand and one finger. The patient lies in the lateral decubitus position facing away from you. With the palm of your cephalad hand against the sacrum at the level of S2/3 to provide counterpressure, place the index or middle finger of the other hand in the rectum. For anterior displacements, push the tip of the coccyx first posteriorly and then upward. For posterior displacements, push anteriorly; for lateral displacements, push from side to side; for fixations, move the coccyx back and forth posteriorly and anteriorly. Gently, slowly and rhythmically increase the pressure. In the beginning, just use the movement to test the mobility and resistance of the tissues. Increase the force slowly to mobilize the joint. This is a technique accomplished not by force but through a delicate touch. If successful, you should distinctly feel the freeing of the fibrosed tissues and the return of sacrococcygeal motion.

With a very old displacement, the sacrococcygeal joint may be ankylosed. In this case, do not try to break the ankylosis by force; this may fracture the joint. Release of the soft tissue laterally will sometimes result in a significant improvement in these cases.

Effects

As you manipulate the coccyx posteriorly, all the supporting tissues in the area are stretched, and often regain their proper tonicity and function. For example, we have successfully treated dozens of cases of chronic urinary tract infection in this way (with laboratory documentation). In view of the unpleasantness of this disease, we believe this technique deserves to be widely known. The coccyx is such an important structure that in our treatment protocol for pelvic disorders it is always manipulated first, followed by the cervix, the sacrum and then whichever other organs are involved.

Manipulation of the coccyx should only be carried out once or twice. If there are no positive results, look elsewhere for the source of the problem. Most often, the source will be the uterus. This is not a region where you should stubbornly attempt to prove that your diagnosis is correct.

The direct effect of manipulation of the coccyx in displacements of the bladder and uterus has been discussed in chapters 9 and 10. But there are other, more distant effects not readily explained. Why does the coccyx have an effect on the kidneys, stomach and other organs? The physiological relationships of the human body are far from being well understood. We do not believe that positive results per se are sufficient; let us all continue our research so that osteopathic medicine and human health can benefit from a more complete understanding of the phenomena we observe.

Afterword

In our profession, it is rare that one is able to clearly and scientifically demonstrate the effects of a technique. Our experiments have been able to show the efficacy of visceral manipulations, but only if the manipulations were precise and appropriate. For example, manipulation of the biliary system in a direction that is even slightly incorrect does not only fail to increase biliary transit, but can stop it altogether.

A bile duct, kidney or any other organ is only able to function harmoniously if its structural environment is free of any constraints. Abnormal tension, restrictions or fibrosis upset the entire physiology. Functional problems and gross disease are likely to follow.

Our hands possess the fantastic privilege of being able to treat disorders. Integrity demands that we increase our knowledge in order to refine our techniques. The study which has produced this book falls far short of enabling us to reach our goal, but it is possible that it will help place us on the right track.

As Rollin Becker, D.O., has so wisely stated: "The tissues alone know."

Bibliography

Barral J-P, Mathieu J-P, Mercier P (1981). *Diagnostic Articulaire Vertébral.* Charleroi: S.B.O.R.T.M.

Contamin R, Bernard P, Ferrieux J (1977). *Gynecologie Generale.* Paris: Vigot.

Cruveilhier J (1852). *Traite d'Anatomie Descriptive.* Paris: Labe.

Delmas A (1975). *Voies et Centres Nerveux, 10th ed.* Paris: Masson.

Fryette HH (1980). *Principles of Osteopathic Technic.* Colorado Springs: American Academy of Osteopathy.

Gregoire R, Oberlin S (1973). *Precis D'Anatomie.* Paris: JB Laillere.

Herman H, Cier JF (1977). *Precis de Physiologie.* Paris: Masson.

Huguier M, Bethoux A (1965). Une technique radiologique d'exploration des prolapsus genitaux et des incontinences d'urine: Le colpocystogramme. *Annales de Radiologie;* 8: 809-828.

Kahle W, Leonhardt H, Platzer W (1978). *Anatomie des Visceres.* Paris: Flammarion.

Kamina P (1984). *Anatomie Gynecologique et Obstetricals.* Paris: Maloine.

Korr I (1978). *The Neurobiologic Mechanisms in Manipulative Theory.* NY: Plenum.

Lansac J, Lecomte P (1981). *Gynecologie Pour le Practicien.* Villourbann: S.I.M.E.P.

Malinas Y, Favier M (1979). *Gynecologie-Obstetrique.* Paris: Masson.

Mitchell FL, Moran PS, Pruzzo NA (1979). *An Evaluation and Treatment Manual of Osteopathic Muscle Energy Procedures.* Valley Park, MO: Mitchell, Moran & Prusso Associates.

Renaud R, Sermont H, Ritter J, Bohler JL, Eborst E, Gamerre M, Jacquetin B, Sormont G (1982). *Les Incontinences Urinaires Chez La Femme.* Paris: Masson.

Robert HG, Palmer R, Boury-Heyler C, Cohen J (1974). *Precis de Gynecologie.* Paris: Masson.

Rouviere H (1967). *Anatomie Humaine.* Paris: Masson.

Scali P, Warrell DW (1980). *Les Prolapsus Vaginaux et l'Incontinence Urinaire Chez les Femmes.* Paris: Masson.

Taurelle R (1980). *Obstetrique.* Paris: France Medical Edition.

Testud L, Latarjet A (1948). *Traite d'Anatomie Humaine.* Paris: Gaston Doin.

Testud L, Jacob O (1922). *Anatomie Topographique.* Paris: Gaston Doin.

Testut L (1889). *Traite d'Anatomie Humaine.* Paris: Octave Doin.

de Tourris H, Henrion R, Delecour M (1979). *Gynecologie et Obstetrique.* Paris: Masson.

Upledger JE, Vredevoogd JD (1983). *Craniosacral Therapy.* Chicago: Eastland Press.

Waligora J, Perlemuter L (1975). *Anatomie.* Paris: Masson.

Williams PL, Warwick R (eds) (1980). *Gray's Anatomy, 36th ed.* Philadelphia: WB Saunders.

List of Illustrations

CHAPTER TWO

CHAPTER FOUR

CHAPTER FIVE

CHAPTER SIX

CHAPTER SEVEN

CHAPTER EIGHT

CHAPTER NINE

CHAPTER TEN

Index

Numbers in boldface refer to illustrations.

A

B

C

D

E

F

G

H

I

J

K

L

P

R

S

T

U

V

W

The Jimmy Fund.

Mrs. John F. Heuchert
4 Wildwood Park Road
Clinton, CT 06413-1807